Table of Contents

Preface xvi

Part 1   Forecasting and Process Analysis 1

Case 1   Forecasting Sales at Ska Brewing Company 3
        Eric Huggins, Fort Lewis College

Case 2   Maintaining Financial Success and Expanding into
        Other Markets at FeedMyPet.com 15
        Charles A. Wood, Duquesne University

Case 3   Forecasting Offertory Revenue at St. Elizabeth Seton
        Catholic Church 25
        Matthew J. Drake, Duquesne University
        Ozgun Caliskan-Demirag, Pennsylvania State University—Erie, The Behrend College

Case 4   Pizza Station 33
        Kathryn Marley, Duquesne University
        Gopesh Anand, University of Illinois at Urbana–Champaign

Part 2   Optimization and Simulation 45

Case 5   Inventory Management at Squirrel Hill Cosmetics 47
        Paul M. Griffin, Pennsylvania State University

Case 6   Safety Stock Planning for a Hong Kong Fashion Retailer ... 65
        Tsan-Ming (Jason) Choi, The Hong Kong Polytechnic University

Case 7   Network Design at Commonwealth Pipeline Company 77
        Matthew J. Drake, Duquesne University

Case 8   Publish or Perish: Scheduling Challenges in the Publishing
        Industry 81
        Beate Klingenberg and David Gavin, Marist College

Part 3   **Decision Analysis** 99

Case 9   Narragansett Brewing Company: Build a Brewery 101
*John K. Visich, Christopher J. Roethlein, and Angela M. Wicks,
Bryant University*

Case 10   Aluminum Versus Plastic: A Life-Cycle Perspective on
the Use of These Materials in Laptop Computers 107
*Ryan Luchs, Drew Lessard, and Robert P. Sroufe,
Duquesne University*

Case 11   HealthCare's Corporate Social Responsibility Program.... 131
*Robert P. Sroufe and Marie Fechik-Kirk,
Duquesne University*

Case 12   PaperbackSwap.com: Got Books? 143
Brandy S. Cannon and Louis A. Le Blanc, Berry College

Case 13   Stranded in the Nyiri Desert: A Group Case Study 161
Aimée A. Kane and Mercy Shitemi, Duquesne University

Part 4   **Advanced Business Analytics** 165

Case 14   Joe's Coin Shop: Entry into Online Auctions 167
Charles A. Wood, Duquesne University

Case 15   Vehicle Routing at Otto's Discount Brigade 181
Matthew J. Drake, Duquesne University

Index 189

The Applied Business Analytics Casebook

The Applied Business Analytics Casebook

Applications in Supply Chain Management, Operations Management, and Operations Research

Matthew J. Drake, Ph.D., CFPIM

Pearson Education, Inc.

Vice President, Publisher: Tim Moore
Associate Publisher and Director of Marketing: Amy Neidlinger
Executive Editor: Jeanne Glasser Levine
Operations Specialist: Jodi Kemper
Cover Designer: Chuti Prasertsith
Managing Editor: Kristy Hart
Project Editor: Katie Matejka
Copy Editor: Seth Kerney
Proofreader: Chuck Hutchinson
Indexer: Johnna VanHoose Dinse
Senior Compositor: Gloria Schurick
Manufacturing Buyer: Dan Uhrig
© 2014 by Matthew J. Drake
Publishing as Pearson
Upper Saddle River, New Jersey 07458

Pearson offers excellent discounts on this book when ordered in quantity for bulk purchases or special sales. For more information, please contact U.S. Corporate and Government Sales, 1-800-382-3419, corpsales@pearsontechgroup.com. For sales outside the U.S., please contact International Sales at international@pearsoned.com.

Company and product names mentioned herein are the trademarks or registered trademarks of their respective owners.

All rights reserved. No part of this book may be reproduced, in any form or by any means, without permission in writing from the publisher.

Printed in the United States of America

First Printing November 2013

ISBN-10: 0-13-340736-5
ISBN-13: 978-0-13-340736-5

Pearson Education LTD.
Pearson Education Australia PTY, Limited.
Pearson Education Singapore, Pte. Ltd.
Pearson Education Asia, Ltd.
Pearson Education Canada, Ltd.
Pearson Educación de Mexico, S.A. de C.V.
Pearson Education—Japan
Pearson Education Malaysia, Pte. Ltd.
Library of Congress Control Number: 2013946942

*For my wife, Nicole, and my daughter, Noelle.
You are the inspiration for
everything that I accomplish.*

Acknowledgments

I am forever grateful to the efforts of all of the contributors to this book. Many of them have been friends and colleagues for a long time, but I met some others for the first time through working on this project. I look forward to many more years of collaboration with them. This book would not have become a reality without the contributors' willingness to share their hard work with me. I am also indebted to Barry Render, Consulting Editor at FT Press, who invited me to work on this project, and to Jeanne Glasser Levine, Executive Editor at FT Press, whose guidance and advice was instrumental throughout the publication process.

About the Author

Matthew J. Drake, Ph.D., CFPIM, is an Associate Professor of Supply Chain Management and the Director of International Business Programs in the Palumbo-Donahue School of Business at Duquesne University. Dr. Drake primarily teaches analytical courses in the Supply Chain Management program. He holds a B.S. in Business Administration from Duquesne University and an M.S. and Ph.D. in Industrial Engineering from the Georgia Institute of Technology. His first book, *Global Supply Chain Management*, was published by Business Expert Press in 2012. Dr. Drake's research has been published in a number of leading journals including *Naval Research Logistics*, the *European Journal of Operational Research*, *Omega*, the *International Journal of Production Economics*, *OR Spectrum*, the *Journal of Business Ethics*, and *Science and Engineering Ethics*. Several of his previous cases and teaching materials have been published in *INFORMS Transactions on Education* and *Spreadsheets in Education*.

Dr. Drake lives in suburban Pittsburgh, Pennsylvania, with his wife, Nicole; his daughter, Noelle; and his dog, Bismarck.

Contributor List

Gopesh Anand is an Associate Professor of Process Management in the College of Business at the University of Illinois at Urbana–Champaign. His research is aimed at understanding continuous improvement of work processes and execution of operations strategy in organizations.

Ozgun Caliskan-Demirag is an Assistant Professor of Supply Chain Management in the Sam and Irene Black School of Business at Penn State Erie, The Behrend College. She holds a Ph.D. in Industrial and Systems Engineering from Georgia Tech, and her main research interests are in the areas of supply chain management, operations/marketing interface, inventory management and decentralized resource allocation. Her work has appeared in journals such as *Operations Research*, *Production and Operations Management*, *Naval Research Logistics*, and *European Journal of Operational Research*.

Brandy S. Cannon is employed as an accountant in the Business and Finance Office at Berry College, Mount Berry, Georgia, USA. She earned a B.S. in Accounting and the M.B.A. from the Campbell School of Business at Berry College.

Tsan-Ming (Jason) Choi is an Associate Professor in Fashion Business at The Hong Kong Polytechnic University. Over the past few years, he has actively participated in a variety of research projects on supply chain management and systems engineering. He has authored/edited 10 research handbooks and published extensively in leading OR/OM journals such as *Annals of Operations Research*, *Automatica*, *Decision Support Systems*, *European Journal of Operational Research*, *IEEE Transactions on Automatic Control*, *Production and Operations Management*, *Service Science*, *Supply Chain Management*, and various other *IEEE Transactions*. He is now an area editor/associate editor/guest editor of journals which include *Annals of Operations Research*; *Decision Sciences*; *Decision Support*

Systems; *European Management Journal*; *IEEE Transactions on Systems, Man, and Cybernetics Part A: Systems*; *Information Sciences*; *Journal of the Operational Research Society*; and *Production and Operations Management*.

Marie Fechik-Kirk, a Fulbright alumnus, earned an M.B.A. with a focus in sustainability at Duquesne University in 2009. Since then she has helped organizations from Bayer MaterialScience to The Hill School in reducing waste, increasing efficiency, and enhancing their reputation through sustainability initiatives.

David Gavin is an Associate Professor of Management at Marist College. He received his doctorate in Strategic Management from the University at Albany. His professional experience includes upper executive positions in the publishing, technology, food service, and retail industries. He has authored or co-authored articles appearing in the *Journal of Business and Economics Studies*, *International Journal of Humanities and Social Science*, and *International Journal of Organization Theory and Behavior*.

Paul M. Griffin is a Professor in the Harold and Inge Marcus Department of Industrial and Manufacturing Engineering, where he serves as the Peter and Angela Dal Pezzo Department Head Chair. His research and teaching interests are in health and supply chain systems. Dr. Griffin earned a Ph.D. in Industrial Engineering from Texas A&M University.

Eric Huggins is an Associate Professor of Management at Fort Lewis College in Durango, Colorado. When he's not busy teaching, working with student, or analyzing data from local companies, he enjoys spending time in the great outdoors of southwestern Colorado, and he can occasionally be found in the tasting room at Ska.

Aimée A. Kane holds a Ph.D. in organizational behavior and theory from the Tepper School of Business at Carnegie Mellon University. She is an Assistant Professor of Management at the Palumbo-Donahue School of Business at Duquesne University. Her research, which focuses on how groups capitalize on the knowledge of their members, has appeared in several top publications, including the *Academy of Management Annals* and *Organization Science*.

Beate Klingenberg is an Associate Professor of Management at Marist College, with a focus on Operations Management and Decision Sciences. Her areas of research include sustainability and environmental management in operations, knowledge management in technology transfer settings, as well as operations management issues in real estate. Her publications appear in academic as well as practitioner publications. Her credentials include a master's in Chemistry and Ph.D. in Physical Chemistry (both University of Erlangen-Nürnberg, Germany) as well as an M.B.A. from Marist College. Furthermore, she has extensive industry experience in technology transfer and project management.

Louis A. Le Blanc is Professor of Business Administration at the Campbell School of Business, Berry College, Mount Berry, Georgia, USA. He received a Ph.D. from Texas A&M University, followed by postdoctoral study at the University of Minnesota and Indiana University. Dr. Le Blanc teaches courses in strategic use of information technology and operations management.

Drew Lessard is a strategy and analytics professional with experience in Global Fortune 500 companies and has a current passion for startups. He holds an M.B.A. concentrating in Sustainability from Duquesne University and a Master of Arts in Economics from Boston University. He hails from Portland, Maine, and currently resides in Pittsburgh, Pennsylvania.

Ryan Luchs is an Assistant Professor of Marketing in the Palumbo-Donahue School of Business at Duquesne University. He teaches marketing and supply chain management courses to undergraduates and also teaches the Strategic Marketing course in the Sustainable M.B.A. curriculum. Dr. Luchs received a Ph.D. and an M.B.A. from the University of Pittsburgh and a B.S. in Chemical Engineering from Penn State University.

Kathryn Marley is an Assistant Professor of Supply Chain Management in the Palumbo-Donahue School of Business at Duquesne University. Her research interests include lean management and continuous improvement programs, supply chain disruptions, and pedagogical methods.

Christopher Roethlein is a Professor in the Management Department at Bryant University where he teaches courses in operations management and supply chain management. He has a Ph.D. in Management Science and Information Systems from the University of Rhode Island; and his research interests include quality and communication within a supply chain, strategic initiatives through alignment of supply chain goals, collaborative relationships, and leadership excellence. He has published in a numerous journals, and he was a co-winner of the 2011 Case Studies Award Competition presented by the Decision Sciences Institute.

Mercy Shitemi holds a B.S. in Informatics from Indiana University and is currently completing a master's degree in Information Systems Management at Duquesne University's John F. Donahue Graduate School of Business. Mercy hails from Eldoret, Kenya.

Robert P. Sroufe is the Murrin Chair of Global Competitiveness in the John F. Donahue Graduate School of Business and Director of Applied Sustainability within the Beard Institute at Duquesne University. Dr. Sroufe is an award-winning scholar and teacher. These awards include instructional innovation and best environmental papers from the National Decision Sciences Institute. Within the M.B.A. Sustainability program, he develops and delivers courses on sustainable theories and models including life-cycle analysis, business applications of sustainability tools, and processes for new initiatives; and he oversees action-learning consulting projects every semester with corporate sponsors.

John K. Visich is a Professor in the Management Department at Bryant University, where he teaches courses in operations management and supply chain management. He has a Ph.D. in Operations Management from the University of Houston, and his research interests are in supply chain management, radio frequency identification, and corporate social responsibility. He has published in a numerous journals, and he was a co-winner of the 2011 Case Studies Award Competition presented by the Decision Sciences Institute.

Angela M. Wicks is an Associate Professor in the Management Department at Bryant University, where she teaches courses in operations management and project management. She has a Ph.D. in Operations Management from the University of Houston, and her research interests include hospital performance, patient satisfaction, and health care technology. She has published in numerous journals including the *International Journal of Quality Assurance in Healthcare*, *Hospital Topics*, and the *International Journal of Healthcare Technology and Management*.

Charles A. Wood is an Assistant Professor in the Management Information Systems area at the Palumbo Donahue School of Business at Duquesne University in Pittsburgh, Pennsylvania. After spending over a decade in the "real world" as a systems analyst, team leader, manager, systems architect, and finally as the owner of a successful consulting company, Chuck returned to academia to complete an M.B.A. and a Ph.D. He has taught at several institutions, including Notre Dame and at the University of Minnesota.

Preface

The field of business analytics has been thrust into the global spotlight in recent years. This surge in popularity is largely because of a barrage of books and periodical articles highlighting its potential to help firms create a competitive advantage. Although some techniques contained within the umbrella of business analytics, such as data mining, text mining, and neural networks, truly represent cutting-edge methodologies that mainly appear in advanced graduate courses, the building-block techniques of business analytics, such as statistical analysis, optimization, and decision trees, are mainstays in business-school curricula around the world.

Business analytics can be broadly defined as "the scientific process of transforming data into insight for better decision making."[1] As a result of this focus on decision making, courses that cover material related to business analytics can benefit greatly from utilizing case studies as a supplement to the core analytical material. Case studies are an effective method for exposing students to the entire decision-making process because they put the student in a simulated active role as a decision maker who must perform the analysis and use the output to recommend a course of action.

Although cases are a mainstay of many graduate business courses, they are used somewhat less frequently in undergraduate courses. One reason for this lack of extensive case adoption in undergraduate courses is the preponderance of long cases published by the major case libraries. Cases appropriate for undergraduates need to be somewhat more focused because the students do not have as much experience as graduate students. Many textbooks include one- or two-page cases at the end of a chapter to illustrate the application of the techniques presented in the chapter. Because they are so short, these cases often amount to little more than a slightly expanded homework problem.

[1] http://www.informs.org/About-INFORMS/What-is-Analytics

This collection of cases is designed to supplement core material covering business analysis techniques in courses as varied as statistics, operations management, management science, supply chain modeling, and decision analysis. This book fills the gap in the library of business analytics case materials appropriate for undergraduate students with cases of moderate length. The cases are also appropriate for introductory-level graduate courses, as instructors can focus the analysis and discussion on more of the complex issues raised in the cases.

The cases in the collection are grouped by the primary analytical technique appropriate for each decision environment. Part 1, "Forecasting and Process Analysis," includes three forecasting cases and one case that focuses on quality control and process improvement. Part 2, "Optimization and Simulation," contains cases that utilize the classic management science methods of optimization and simulation. The optimization cases address inventory control and logistics network design, and the simulation case addresses the management of process flows. Part 3, "Decision Analysis," includes cases that require the application of a variety of decision analysis tools from decision trees and factor rating to the Analytic Hierarchy Process (AHP), multi-criteria decision analysis, and group decision making. The decision environments vary from facility location to sustainability management. Part 4, "Advanced Business Analytics," contains two advanced cases—one that is truly a "big data" case with a large data set and another centered on vehicle routing, a traditionally difficult problem in logistics.

It is my hope that the cases in this collection expose students to the power of business analytics and the utility of these techniques in the decision-making process. Students armed with an effective toolbox of analytical skills and techniques are well positioned to make thoughtful, reasoned decisions informed by data analysis for their companies and organizations. These analytical skills are transferrable across companies and industries and can enhance students' attractiveness and value to employers throughout their careers.

Matthew J. Drake
Pittsburgh, Pennsylvania, USA
August 2013

1
Forecasting and Process Analysis

1. Forecasting Sales at Ska Brewing Company   3
2. Maintaining Financial Success and Expanding into Other Markets at FeedMyPet.com   15
3. Forecasting Offertory Revenue at St. Elizabeth Seton Catholic Church   25
4. Pizza Station   33

Case 1

Forecasting Sales at Ska Brewing Company

Eric Huggins, Fort Lewis College

Background

Ska Brewing Company is a purveyor of fine craft beers located in Durango, Colorado. With its flagships Pinstripe Red Ale and True Blonde Ale, medal-winning Buster Nut Brown Ale and Steel Toe Stout, and seasonal Mexican Logger and Euphoria Pale Ale, Ska has enjoyed double-digit growth for more than a decade with no signs of slowing down. Learn more about Ska by visiting its tasting room at 225 Girard Street, Durango, Colorado, or online.[1]

In the early '90s, founders/owners Dave and Bill were dissatisfied with watered-down corporate beer and decided to take matters into their own hands, literally. They began brewing their own beer in their basement, much to the delight of everyone who knew them. Eventually, it became clear that they might be able to make a living doing what they loved to do, and they founded Ska Brewing Company in 1995 with third owner/founder Matt. Through hard work and a laser-like focus on brewing great beer, Ska continued to grow, and in 2008

---

[1] http://www.skabrewing.com/

the company moved into its $4.8 million, 24,000-square-foot world headquarters. In 2012, Ska brewed more than 25,000 barrels of beer (1 barrel = 2 standard kegs = 252 pints = 4,032 ounces), with sales exceeding $6.5 million.

Ska was not alone in its success. Durango, a town with fewer than 20,000 people, has four long-term successful breweries/brewpubs, a brand new brewpub that opened in 2012, and another one in the works. Rather than considering these other breweries as competition, Ska has worked together with them (as well as others across the state of Colorado) to brew specialty beers for festivals and other occasions; Ska also contract brews beer for Steamworks Brewing Company (using its recipes) because Steamworks has exceeded its own brewing capacity. Owner Dave calls this unique relationship "coopitition." Steamworks and Ska are just examples, however.

The craft brewing industry has seen phenomenal growth during the last three decades across the United States and in other countries as well. According to the Brewers Association,[2] the craft brewing renaissance started in the late 1970s and saw periods of incredible growth during the 1990s. Historically, before Prohibition, small breweries were everywhere across the United States; the 18th Amendment caused most of the small breweries to go out of business, and only the larger breweries survived until the 21st Amendment repealed Prohibition 13 years later. It took several decades for smaller breweries to begin the resurgence that we see today.

But our concern is more specific: Will the growth and success at Ska continue? Can Ska anticipate how much beer it will produce, and what sales will be so that the company can plan wisely for the future? In fact, current plans are to increase brewing capacity yet again—a costly investment with potentially high returns. Is this a good decision or not? This is where *you* come in.

---

[2] http://www.brewersassociation.org/pages/about-us/history-of-craft-brewing

Mission

Despite its success, Ska is still a relatively small operation. The company has one main numbers person, accountant Erik. In a nutshell, Erik would like to predict Ska's sales dollars and barrels sold for the current year, 2013. He has done some of this work on his own, but he would like you to confirm (or refute) his forecasts, and to do so in much greater detail, as Erik is too busy (presumably because he spends his days counting all of Ska's money). To get you started, Exhibit 1.1 contains Ska's total barrels (BBLS) sold and sales ($$$) over the previous 13 years. More precise monthly data is available in Exhibits 1.2 through 1.6. Please note that this is actual (not phony textbook) data.

Even a cursory glance at the information in the table shows that both the number of barrels and sales are increasing annually at a pretty good rate. In fact, both values have shown tenfold growth between the years 2000 and 2012. What will these two numbers look like at the end of 2013? You might have studied forecasting techniques previously, and ideally you learned that when forecasting real data, there is no "one-size-fits-all" approach; ahead you will try several approaches and then combine them to make a final prediction.

Your task is not only to forecast these two values for 2013, but to give Erik, Dave, Bill, and Matt a better picture of what is happening with their business overall. To do so, you will be asked to produce several graphs, both on annual and monthly bases, to consider growth as a percentage, and to consider the likely errors that go along with your forecasts. You will first be asked to learn a little more about the brewing industry in general, to give you a better idea of the current status of craft brewing. Your final report should be thorough, professional, and accurate. Good luck!

Questions about Breweries

1. What is a craft brewery? How is it different from a brewpub? Go online and research these definitions. You should fairly easily find a quantitative definition of the number of barrels produced by a craft brewery (or microbrewery). For comparison, find out how many barrels are produced annually by a very large brewery such as Anheuser-Busch, MillerCoors, or Heineken. Write a paragraph or two with your findings and, as always, be sure to cite your sources.

2. Are there any local breweries in your area? If so, which categories do they fall under? If not, why not? Discuss the feasibility and likely success or failure of a new brewery in your area. Of course, a cool name like Ska might be one of the keys to a new brewery's success; what will you name your new brewery?

3. The claim was made earlier that the "craft brewing industry has seen phenomenal growth during the last three decades." Go online and find evidence to support this claim. Specifically, how many craft breweries are there now compared to 30 years ago? How has the craft brewing market share grown (out of total beer sales)? How have the major breweries reacted to the growth of craft brewing? Write a paragraph or two with what you learn.

Questions about Ska's Annual Data

4. Now onto Ska's annual data: Use Microsoft Excel to draw scatter plots of both year versus barrels and year versus sales. (Hint: You might want to change the year range from 2000–2012 to 0–12 to simplify the equations of the curves that Excel will eventually fit to the data.) What kind of curve do both scatter plots look like? Consider the barrels data first; then repeat for the sales data:

a. Have Excel fit a linear trendline to the data and determine the equation of the line and the r^2 value. Interpret the slope of the line and the coefficient of determination. Is this a good fit?

b. The pattern on the graph should be clearly nonlinear. Now instead, have Excel fit an exponential curve to the data and again determine the equation of the curve and the r^2 value. Is this a better fit?

c. Using the equation for the curve from 4b, plug in 13 (or 2013) to get your first forecast. Does it seem reasonable, or does it seem too low or too high? (Note: To see where the forecast falls, Excel will let you extend the curve by one period when you draw the trendline. When you format the trendline, forecast forward one period.)

Be sure to do 4a–4c for both barrels and sales.

5. Now draw a scatter diagram of barrels versus sales. This pattern should appear quite linear. Fit a line to the data and interpret both the slope of the line (Hint: 1 barrel = 2 kegs) and the coefficient of determination. Can you reasonably conclude that the more beer Ska produces, the more money it makes?

6. Reconsider the graphs from question 4. Although the growth does appear to be exponential, your predictions in 4c shouldn't quite look right. Let's try it another way: Consider the last four points on each graph, from 2009 to 2012. Ignoring the rest of the data, do those four points appear to have an (obvious) pattern?

    a. Using only the last four years' data, fit a line for both barrels and for sales. Interpret both the slope and r^2 value for each line.

    b. Plug a 13 into each line to get your second forecast for barrels and sales in 2013. How confident do you feel with these predictions?

7. Your predictions in question 6 might seem pretty good, but take it one step further:

   a. For both barrels and sales, determine the MAD for each of your predictions. If you are not familiar with the concept of MAD, go online and search for "mean absolute deviation." You should quickly find a website that explains the concept and shows you how to calculate it. What are your forecasts for 2013 including the MAD? What information does the MAD tell you?

   b. Repeat 7a but now for the MAPE, or mean absolute percentage error. Interpret the MAPE.

   c. As one final check, repeat what you did in question 6 but this time use the data from 2008 to 2011 to predict 2012 and compare your prediction for 2012 to the actual value. Do this for both barrels and sales. Does this forecasting method appear to be promising?

8. In both 4c and 6b, you forecasted barrels and sales for 2013. Consider one more way to do this before you make your final decision. Determine the percentage growth for both barrels and sales for each year. For example, from 2000 to 2001, barrels increased from 2,595 to 3,025, or a growth rate of (3025 − 2595)/2595 = 17%. Calculate these rates for years 1 to 12 for both columns of data.

   a. Determine the average and median growth rates for both barrels and sales.

   b. Considering only sales, draw a scatter plot of year versus sales growth. Do any of the growth rates look like outliers? (Hint: Recall that Ska moved into its new world headquarters in 2008, increasing its brewing capacity tremendously.)

   c. The outliers in 8b might be obvious, but they aren't always so easy to identify. So, use a box plot (Tukey's Method) to find the outliers. For each column of percentage data, determine the first and third quartiles; these are the points where 25% of the data are below and 25% of the data are above, respectively. (Hint: Use Excel's =quartile() function to find both Q1 and Q3.) Calculate the IQR = Q3 − Q1 and the range of "typical" values (Q1 − 1.5°IQR, Q3 + 1.5°IQR). Any data point within the range is typical, whereas any point outside the range is atypical, or an outlier. What are the two outliers for each column in this case?

d. Eliminate the outliers and recalculate the average and median growth rates for both barrels and sales. Multiply these growth rates by the 2012 actual values for barrels and sales and make your third (and *final*) set of forecasts for 2013. How do you feel about these predictions?

e. As a side note, Erik, the accountant, asked the owners to do a quick, back-of-the-beer-coaster estimate of what growth would be for 2013. Their immediate response was "20%." Would you say that Dave, Bill, and Matt are guessing, or do they know their business very well?

Questions about Ska's Monthly Data

Another concern at Ska is seasonal variation. The brewery is much busier during the summer months than during the winter months. Two possible explanations for this phenomenon are that 1) people simply buy more beer during the summer, and 2) Ska releases two very popular seasonal beers, Mexican Logger and Euphoria Pale Ale, at the beginning and end of the summer season. To get a better handle on the seasonal variations at Ska, your task is to draw some clear pictures of what's happening (sometimes called *data visualization*).

To achieve this goal, consider Exhibits 1.2–1.6 with the complete monthly data for all 13 years. You will see the barrels information in white and the sales information in gray. Use this data to display the seasonal patterns at Ska:

9. Thirteen years provides 156 months' worth of data. In Excel, develop one column from 1 to 156. In the next column, list the barrels sold for each year in chronological order (so the first 12 data points will be the 196.5–238.1 from year 2000, the next 12 will be the 243.2–258.9 from year 2001, and so on.) (Hint: You can build this using simple cut/copy and paste, or there's likely a better way.) In the third column, list all the monthly sales data.

   a. Graph a scatter plot of both month versus barrels and month versus sales.

   b. Fit exponential curves to both graphs.

   c. Look carefully at the last four years of each graph. When does Ska tend to get busier during these four years? Does each graph indicate that summertime is crunch time? Which months in particular appear to be the busiest?

10. For the final forecasts for 2013, predict each month of 2013 and add them to the scatter plot from question 9. As you did in question 4, use only the last four years from 2009 to 2012 to forecast 2013.

    a. For each month, make a linear forecast using the monthly data from 2009 to 2012. So, for example, to predict barrels for January 2013, use the data points 706.6, 1017.3, 1272.4, and 1484.9, and make a straightforward linear prediction. Do this for both barrels and sales for each month.

    b. Now, add these forecasted values onto the scatter plots from question 9. Make the forecasted values a different color from the actual data to make them stand out and label the final graphs accordingly. These two graphs should give the stakeholders at Ska a clear picture of what 2013 might look like, depending on how accurate the forecasts end up being.

(Note: Adding these extra points to a pre-existing scatter plot in Excel is a little tricky. To do so, right-click on the scatter plot itself and choose Select Data. Click the Add button and add your forecasted values as a new series of data.) According to the two graphs (including actual monthly data and forecasted values), when will Ska be busiest in 2013?

Conclusion

Congratulations, you have just completed a very thorough analysis of Ska Brewing Company's production in barrels and sales figures. At this point, it might be worth reconsidering how accurate forecasts will help Ska. According to Erik, "An accurate sales budget is the root of the entire budgeting process." In addition, Dave says that accurate forecasts would help "tremendously," allowing Ska to "increase efficiencies from a production standpoint," and help "make decisions about whether or not Ska could enter any new markets."

Now it's time to tie everything together and make your best forecast for 2013 for both barrels and sales, including some kind of estimate of the error term. Carefully combine your forecasts from 4c, 6b, 7a, 7b, and 8d. Be bold and use a large font—*you* are an expert now!

Year	Forecasted Barrels	Forecasted Sales
2013		

(Note to students: The actual values for 2013 have not yet been realized as I (the author) prepare this case study. When they become available in early 2014, I will get them from Ska and record them. If you are curious about how good your final forecasts actually were, send them to Dr. Eric Huggins,[3] and I'll reply with the actual values when they become available.)

[3] huggins_e@fortlewis.edu

Exhibits

Exhibit 1.1 Barrels Sold and Sales Volume at Ska Brewing Company

Year	BBLS	$$$
2000	2,595	$521,050
2001	3,025	$629,866
2002	3,465	$739,153
2003	4,031	$883,378
2004	4,525	$1,011,409
2005	5,273	$1,234,628
2006	6,268	$1,481,759
2007	7,289	$1,754,272
2008	7,943	$2,080,795
2009	11,681	$3,179,390
2010	16,026	$4,376,982
2011	21,258	$5,317,535
2012	25,771	$6,553,145

Exhibit 1.2 Monthly Data for Barrels and Sales (2000–2002)

	2000		2001		2002	
	BBLS	$$$	BBLS	$$$	BBLS	$$$
Jan	196.5	$40,458	243.2	$53,093	290.80	$62,989
Feb	193.2	$35,615	239.9	$48,819	254.80	$53,912
Mar	229.7	$43,306	241.3	$49,782	267.50	$56,477
Apr	190.2	$34,885	214.2	$44,515	252.30	$54,720
May	195.1	$40,879	227.6	$50,671	306.90	$67,387
Jun	261.9	$53,378	309.3	$64,764	323.80	$68,196
Jul	230.2	$46,850	292.5	$59,947	336.80	$71,179
Aug	247.7	$50,118	327.9	$67,821	326.00	$69,643
Sep	210.6	$43,872	226.2	$46,102	272.70	$57,426
Oct	203.3	$41,805	242.9	$50,403	262.90	$55,944
Nov	198.2	$39,317	201.3	$40,892	248.70	$55,639
Dec	238.1	$50,569	258.9	$53,058	321.70	$65,643
Total	2,594.7	$521,050	3,025.2	$629,866	3,464.9	$739,153

Exhibit 1.3 Monthly Data for Barrels and Sales (2003–2005)

	2003		2004		2005	
	BBLS	$$$	BBLS	$$$	BBLS	$$$
Jan	336.3	$74,218	336.6	$74,072	411.1	$93,769
Feb	269.4	$57,567	317.6	$69,847	374.7	$83,907
Mar	284.8	$58,822	405.6	$86,334	396.2	$95,687
Apr	273.5	$59,089	344	$76,337	388.4	$91,712
May	367.0	$81,849	391.5	$94,853	435.9	$101,683
Jun	383.0	$85,636	492.1	$110,477	526.3	$122,572
Jul	388.3	$87,031	410.9	$90,578	492.1	$114,097
Aug	416.1	$91,560	418.4	$90,768	492.7	$114,925
Sep	292.5	$65,976	411.3	$90,308	449.6	$103,899
Oct	386.0	$84,438	309.6	$69,961	434.3	$104,706
Nov	266.0	$58,368	308.7	$69,073	433.4	$102,354
Dec	368.1	$78,824	378.9	$88,800	437.9	$105,317
Total	4031	$883,378	4525.2	$1,011,409	5272.6	$1,234,628

Exhibit 1.4 Monthly Data for Barrels and Sales (2006–2008)

	2006		2007		2008	
	BBLS	$$$	BBLS	$$$	BBLS	$$$
Jan	455.9	$107,422	598.7	$141,177	581.5	$153,098
Feb	437.2	$101,485	512.4	$124,511	628.7	$163,893
Mar	619.7	$140,082	560.3	$133,152	658.3	$164,180
Apr	368.6	$88,973	628.5	$142,942	628.5	$176,973
May	635.2	$149,576	621.9	$151,621	685.7	$177,043
Jun	587.5	$139,916	780.1	$182,735	661.8	$168,823
Jul	597.8	$141,982	641.9	$152,912	780.8	$201,482
Aug	557.1	$133,007	728.6	$176,702	725.5	$190,317
Sep	567.6	$132,330	571.3	$136,517	626.8	$156,337
Oct	478.3	$115,470	641.2	$159,959	676	$186,388
Nov	424.1	$100,695	418.8	$107,104	518.5	$138,374
Dec	538.8	$130,823	585	$144,939	770.9	$203,889
Total	6267.8	$1,481,759	7288.7	$1,754,272	7943	$2,080,795

Exhibit 1.5 Monthly Data for Barrels and Sales (2009–2011)

	2009		2010		2011	
	BBLS	$$$	BBLS	$$$	BBLS	$$$
Jan	706.6	$193,481	1,017.3	$267,782	1,272.4	$319,313
Feb	641.3	$170,674	853.3	$225,592	1,275.9	$323,726
Mar	884.8	$228,095	1,124.2	$356,604	1,333.2	$342,353
Apr	862.4	$232,372	999.1	$274,723	1,356	$361,315
May	1,061.3	$288,188	1,434.1	$377,369	2,471.1	$612,500
Jun	1,110	$304,763	1,673.3	$439,907	2,276.3	$564,599
Jul	1,269.5	$337,825	1,626.7	$430,999	2,102.3	$518,422
Aug	1,269.7	$342,121	1,871.7	$485,822	2,556.2	$623,860
Sep	1,147.5	$320,011	1,398	$407,577	1,631.4	$412,091
Oct	1,107.4	$304,756	1,649.4	$450,234	2,140.4	$530,636
Nov	766.1	$221,514	1,111.2	$315,238	1,258.1	$313,034
Dec	854.8	$235,591	1,267.5	$345,135	1,584.3	$395,686
Total	11,681.4	$3,179,390	16,025.8	$4,376,982	21,257.6	$5,317,535

Exhibit 1.6 Monthly Data for Barrels and Sales (2012)

	2012	
	BBLS	$$$
Jan	1,484.9	$375,117
Feb	1,520.9	$391,677
Mar	1,624.2	$426,746
Apr	2,136.1	$535,876
May	2,622.2	$659,204
Jun	2,349.6	$582,670
Jul	2,635	$663,534
Aug	2,292.9	$564,901
Sep	2,495.2	$636,399
Oct	2,856.7	$727,822
Nov	2,088.3	$539,011
Dec	1,664.7	$450,188
Total	25,770.7	$6,553,145

Case 2

Maintaining Financial Success and Expanding into Other Markets at FeedMyPet.com

Charles A. Wood, Duquesne University

Introduction

The first-ever 10-Q quarterly financial reports have just been filed with the SEC (U.S. Securities and Exchange Commission), and John McCloud is very happy with his company's performance so far. FeedMyPet.com just conducted its first IPO last month, and has raised an amazing $89 million after only one year in business (see Exhibit 2.1). After starting FeedMyPet.com last February, just a little over a year ago, Cindy Jones, FeedMyPet.com's COO (Chief Operating Operator), joined John McCloud, founder and CEO (Chief Executive Officer) of FeedMyPet.com, to review the year's company activities.

Like many new startups, FeedMyPet.com had a rough time getting started. Expenditures were high, especially in the area of advertising, which was necessary to increase name recognition. There were also some problems with an inadequate business plan formulation and a lack of initial market research that caused some industry analysts ("negativos" as McCloud calls them) to be unenthusiastic about the company, but the market has spoken, and McCloud couldn't help but smile as he thought of the investing community—the true visionaries in the business world—who flocked to the recently held IPO. The additional capital will allow the firm to accomplish amazing things.

FeedMyPet.com is an online business that sells pet food, pet accessories, and pet supplies direct to consumers over the Internet. The public offering raised so much money that FeedMyPet.com was able to purchase its main competitor, KennelTime.com, leaving it to rule the entire online pet marketplace. McCloud attributes his vast success to several factors:

- FeedMyPet.com has a dedicated support staff of 280 people (whereas most online pet shops have around 30 employees). This staff is much more responsive to customer needs than the skeleton staff at other online pet product companies.

- FeedMyPet.com offers free shipping. Dog food bags and cans are heavy, and the cost of shipping can approach the cost of the dog food. Some competitors have tried to pass these costs on to the consumer, thus alienating the client base; but FeedMyPet.com did not make that mistake and has been rewarded with the largest market share of any online pet product retailer.

- FeedMyPet.com has an advertising campaign that spans across a variety of media, including TV, print, radio, web-based ads, and even its own FeedMyPet.com magazine. This marketing was important, especially right before the IPO, when advertising drove up the price of the initial stock!

The advertising campaign started last year with a 5-city advertising campaign rollout and has now expanded to 10 cities, and has finally gone nationwide with a $1.4 million Super Bowl ad that introduced the country to its answer as to why customers should shop at FeedMyPet.com: "Because Pets are People Too!" The cute ad featured a large man in a dog costume, won several awards, and had the highest recall of any ad that ran during the Super Bowl. Name recognition was at an all-time high. After the ad, FeedMyPet.com went public with an IPO that raised millions. McCloud is sure that extensive advertising has played a large part in FeedMyPet.com's success and that the advertising budget will continue to grow to ensure that the company remains successful.

To deliver pet supplies, FeedMyPet.com made significant investments in infrastructure such as computer networking and data warehousing. FeedMyPet.com's management maintained that the company needed to realize a revenue run rate that supported this infrastructure build-out. McCloud's fellow executives believe that revenue needs to approach $400 million to hit the break-even point, and that it will take a minimum of four to five years to hit that run rate. But investors are still on board; with the stock price on the way up, McCloud perceives no problem with cash flow at FeedMyPet.com.

As McCloud and Jones contemplate the future, they discuss two important topics that require Jones to conduct some additional investigation:

- **Maintaining financial success**—Clearly, FeedMyPet.com has become the de facto leader in the online pet products industry in a fairly short time. Investment and revenues have followed this market leadership. How might FeedMyPet.com's leadership leverage into other ventures? And how does FeedMyPet.com not only dissuade future online competitors, but also coerce current offline "bricks and mortar" shoppers to move their shopping for pet products online?
- **New marketing plan**—Marketing has clearly been the key to FeedMyPet.com's success. The Super Bowl ad drove up investment, and all the previous marketing campaigns have increased market share. Now that FeedMyPet.com is flush with capital, additional marketing plans should be considered.

With so much available cash, McCloud feels comfortable in attacking these issues. McCloud and Jones have discussed various responses to these topics, and the $89 million that they have raised in the recent IPO will fund strategies that heretofore have not been possible.

Maintaining Financial Success

Jones's first task is to ensure FeedMyPet.com maintains its strong financial position. Both McCloud and Jones agree that FeedMyPet.com is in an excellent financial position, especially after the last infusion of $89 million in capital from the recent IPO. McCloud knows that there is a reason for the investor exuberance. Exhibit 2.5 (taken from data found in Exhibit 2.3) graphically shows the explosive growth in sales since the company's inception in the second quarter of last year.

The growth in revenue has been a boon to the company. Total assets have increased 25% from the second quarter to the third quarter. Exhibit 2.6 (taken from data found in Exhibit 2.2) shows the increase in asset value of the company during the same period.

McCloud and Jones discuss how the explosive growth in revenue has sent a strong message to the investor community that FeedMyPet.com is not just a niche small company, but a major player in online retail, and a safe yet profitable investment.

In addition, there are some challenges when operating in the pet supplies industry. Dog food is heavy and costly to ship. Also, to gain market share and name recognition before competitors could swoop into this lucrative market, FeedMyPet.com was very aggressive in the initial pricing of its products, paying $16 million for goods sold to customers for $7 million. This means that in that initial period, for every dollar that FeedMyPet.com paid employees, pet food manufacturers such as Purina, and delivery services such as UPS, it charged the customer about 44 cents. *BusinessWeek*[1] notes that operating margins for pet products retailers are typically much higher; for example, offline "bricks and mortar" pet supplies companies such as Petco typically post a profit margin of up to 4.5%. Jones is confident that after the convenience of online pet delivery catches on, FeedMyPet.com will

---

[1] Arlene Weintraub and Robert D. Hof, "For Online Pet Stores, It's Dog-Eat-Dog," *BusinessWeek*, March 6, 2000.

be able to command an even higher 20%–30% operating margin. Customers are very loyal, and appreciate having pet products, like dog toys and bird food, shipped directly to them rather than forcing the customers to visit a store.

The question remains as to how to best ensure that the success of FeedMyPet.com continues into the future. McCloud and Jones both understand that weak financial positions can drive down stock price and can be exploited by new entrants into the lucrative online pet products market space. Jones's first task is to examine the financial statements (found in the exhibits) to examine how to continue this strong performance into the foreseeable future.

New Marketing Plan

Jones's second task is to formulate a new marketing plan that picks up from the highly successful previous marketing plan. McCloud believes that marketing was the key to FeedMyPet.com's initial success, and Jones is expected to devise a plan to continue that marketing success...no small feat, to be sure! There are several advertising mediums that are available to FeedMyPet.com:

- **TV advertising**—This is the most expensive type of advertising, but the amazing results with the Super Bowl ad show its effectiveness.
- **Magazine/newspaper**—Magazines ads are one of the most targeted advertising mediums. For example, an advertiser can select a pet magazine in which to advertise. Newspapers facilitate geographic targeting.
- **Direct mail**—Direct mail allows a company to target an individual, regardless of online access. It is often inexpensive, and a company can purchase a potential customer list based on demographic information.

- **Telemarketing**—Telemarketing requires a temporary service or employees, but allows a script to be delivered to the target over the phone lines. However, some states enforce a do-not-call list that might interfere with the marketing effort.
- **Search engine optimization (SEO)**—This is probably the most cost-effective method of online advertising. With SEO, you design your web pages and links to optimize their appearance within a web page within the "free" listings, which are then often clicked by the end user.
- **Pay per click (PPC)**—PPC advertising is when search engines charge companies for each click. The bright side of this advertising is that you only pay for the clicks you receive. The downside is that many end users avoid clicking on advertisements.
- **Email marketing**—Email marketing, often called *spamming*, is reviled by many recipients. However, the positive side is that you can easily send out millions of emails for free after you have the email addresses, and a very small percentage of conversions can result in a large increase in sales.

Jones is aware of the various costs of each advertising medium (see Exhibit 2.4). FeedMyPet.com has proven itself willing to spend quite a lot for advertising. Exhibit 2.7 shows how much FeedMyPet.com has spent on advertising since its inception; although, a leveling off has been observed in the last two quarters, perhaps indicating an optimal advertising level for the company.

Final Thoughts on Analysis

Jones has to hit the ground running on these two goals. First, she needs to isolate any potential problem areas in the financial statements. With $89 million, the company is sure to be set for years to come, and it will benefit the company to leverage its new revenue to make the company even more profitable and competitive. Second, she has to develop a marketing plan that allows the company to continue its upward path. Perhaps viewing the numbers as a percentage of sales or developing a nonlinear trend line in Excel to project future financial statement values is in order.

Exhibits

Exhibit 2.1 FeedMyPet.com IPO Capital Raised Statement

	Price Per Share	Shares	Total Revenue
Public offering price	$12.00	8,000,000	$96,000,000
Underwriting discount	$0.88	8,000,000	$7,040,000
Proceeds, before expenses, to FeedMyPet.com	$11.12	8,000,000	$88,960,000

Exhibit 2.2 Quarterly FeedMyPet.com Balance Sheet Data

	Last Year	This Year	
	Third Quarter	Fourth Quarter	First Quarter
Cash and cash equivalents	$52,172	$43,482	$84,137
Current assets	$54,275	$51,967	$110,096
Total assets	$69,695	$86,846	$137,750
Current liabilities	$8,934	$10,903	$16,402
Total liabilities	$10,231	$11,028	$17,376
Total stockholders' equity, including convertible preferred stock	$59,464	$75,818	$120,374

Exhibit 2.3 FeedMyPet.com Statement of Operations (in Thousands)

	Last Year			This Year
	Second Quarter	Third Quarter	Fourth Quarter	First Quarter
Net sales	$47	$682	$6,202	$9,181
Cost of goods sold	($91)	($2,119)	($13,884)	($15,018)
Gross profit	($44)	($1,438)	($7,682)	($5,837)
Operating expenses:				
Marketing and sales	1,346	12,832	36,811	34,626
Product development	1,949	2,633	3,175	3,223
General and administrative	1006	1,446	2,653	2,797
Amortization of stock-based compensation	--	1,367	1174.8	1,290
Total operating expenses	($4,301)	($18,277)	($43,814)	($41,936)
Operating profit (EBIT)	($4,345)	($19,715)	($51,497)	($47,773)
Interest income	$148	$692	$589	$868
Net income	($4,198)	($19,022)	($50,908)	($46,906)
Outstanding shares	1,744	1,744	1,760	9,760
Earnings per share (EPS) (in dollars)	($2.41)	($10.91)	($28.92)	($4.81)

Exhibit 2.4 Advertising Medium Costs[2]

	Setup Process	Setup Cost	Cost of Media
National TV ad spot	Design + production	$50,000–$750,000	$35,000 to $2 million per 30-second spot
National magazine	Design	$1,500–$20,000	$3,000–$25,000 per full-page ad per issue
National newspaper ad	Design	$1,500–$20,000	~$28,000 per half-page ad per day
Direct mail	Design	$1,500–$15,000	~$2.20 per addressee

[2] From WebPageFX, 2009, last accessed Feb. 26, 2013, available at http://www.webpagefx.com/blog/business-advice/the-cost-of-advertising-nationally-broken-down-by-medium/.

	Setup Process	Setup Cost	Cost of Media
Telemarketing	Script writing	$1,000–$4,000	$20–$60 per hour per outbound caller
National SEO	Website configuration	$4,000–$10,000	~$500/month to Internet marketer
National PPC	Campaign configuration	$4,000–$10,000	5¢–$3 per qualified visitor
National email marketing	Email template design	$4,000–$10,000	~$500/month to Internet marketer

Exhibit 2.5 Net Sales Quarter by Quarter

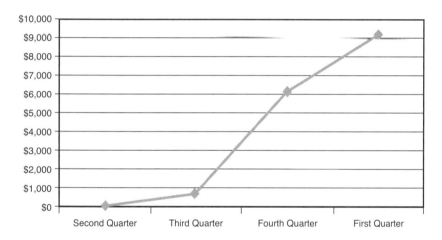

Exhibit 2.6 Asset Growth in the Previous Three Quarters

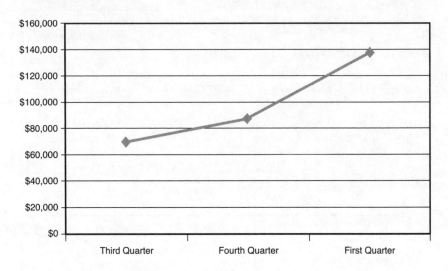

Exhibit 2.7 FeedMyPet.com Quarterly Expenditures on Advertising

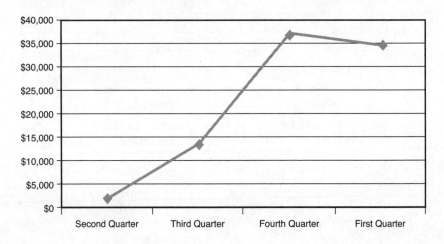

Case 3

Forecasting Offertory Revenue at St. Elizabeth Seton Catholic Church[*]

Matthew J. Drake, Duquesne University

Ozgun Caliskan-Demirag, Pennsylvania State University—Erie, The Behrend College

Introduction

Fr. Clyde Jarreau could not sleep early in the morning of October 10, 2005. The evening before, he had presided over his parish's monthly finance committee meeting, where concerned parishioners examined the church's monthly financial statements and provided recommendations to keep the organization on track financially. At the previous night's meeting, a few of the committee members continued to voice their concern that spending was out of control. The church's bank account balances had fallen sharply for the sixth month in a row, and the committee members were worried that the church would run out of funds sometime early in 2006.

Fr. Jarreau appreciated their commitment to the parish, but he did not need them to remind him of the church's financial struggles. As the pastor of the church, he was greeted by the stack of unpaid bills dominating his desk every time he entered his office. He also saw the

---

[*]Finalist in the 2011 INFORMS Case Competition

stagnant, if not dwindling, weekly offertory collection figures that he had to publish in the weekly bulletin. With expenses increasing without the additional revenue from collections to cover them, Fr. Jarreau knew that he would have many more sleepless nights if he could not find a way for the church to live within its financial means.

After contemplating the problem over a cup of coffee in the rectory's kitchen, Fr. Jarreau knew that he could not construct a solution to such a big problem by himself. He decided to place a call in the morning to a few of his most trusted advisors on the finance committee. These people had been parishioners at the church for more than 15 years, predating himself by a half dozen or so years. They knew the history of the parish over this time and had seen the financial position deteriorate over the past few years, as well. Fr. Jarreau knew that they would do anything they could to help the parish; he only hoped that he was not reaching out to them too late.

St. Elizabeth Seton Catholic Church

Fr. Jarreau's parish, St. Elizabeth Seton, was founded in 1976 in Daphne, Alabama, a small Gulf town just outside of Mobile. The parish grew rapidly throughout the 1980s and 1990s as a large number of workers from the northern United States moved into the south, chasing both displaced jobs and better weather. Although the church itself is the same size as the original building built in 1976, the parish conducted two successful capital campaigns in the subsequent decades after the parish was founded. The first campaign, kicked off in 1985, raised funds to build an educational building for religious education classes for children and adults. The second campaign, begun in 1996, enabled the church to build new offices for its staff and parishioner organizations.

By the year 2000, membership in the church was strong, and the cash reserves were rising each month as parishioners gave generously each week. With the crash of the dot-com bubble in late 2001, however, offertory revenue slid in 2002, and a few years passed before it showed signs of any significant recovery. In an effort to revitalize the church and in keeping with the historical 10-year cycle, Fr. Jarreau spearheaded a new capital campaign toward the end of 2004 with the goal of raising money to build a new recreational hall for the church. This would enable the parish to hold more fellowship activities, as well as generate additional sources of revenue by hosting wedding receptions and other banquets. Unfortunately for the pastor, these new revenue streams would only begin after the building was completed in early 2007.

When Fr. Jarreau initially discussed the new capital campaign in the summer of 2004, several members of the finance committee were worried that many parishioners would simply direct a large portion of their weekly offertory contribution to the new capital campaign. This would severely hinder the church's ability to meet its normal operating expenses. Luckily, however, Fr. Jarreau's explanation of the capital campaign had largely convinced the parishioners to support it in addition to maintaining their normal weekly offertory contributions. The offertory figures thus far in 2005 appeared to be unaffected by the capital campaign.

Cash Flow Analysis

Fr. Jarreau was able to arrange a meeting with his trusted finance committee members, John Gust and Charlie Stewart, a few nights later. When they arrived in his office around 7 p.m., Fr. Jarreau wasted no time summarizing the problem facing the church. "Our bank balances have consistently fallen throughout this year. At this rate, it looks like the parish is going to be out of money by this time next year. What do you guys think we should do?" Charlie and John had thought that they were going to have to open Fr. Jarreau's eyes to the church's financial problems in this meeting, but now it was obvious that the pastor understood them all too well. After a brief sigh of relief, Charlie started, "John and I have been members of the finance committee for years, and we've got a lot of ourselves invested in this parish. We've been trying to suggest subtly that the church's spending was getting out of control, but now it appears as if the time for subtlety has passed. We need to drastically rein in expenses."

Fr. Jarreau recalled their previous concerns but could not reconcile one aspect of the church's financial operations. "But the Archdiocese requires that we have a balanced budget each year. They won't accept a budget from us that isn't balanced. How could we be in this situation with a balanced budget?"

John chimed in, "Well, Father, the problem seems to be with the budget process itself. In my opinion, we've been doing the whole thing backwards. We have been asking each department head to submit his or her expected expenses for the upcoming year, and we have most of our discussions as a committee about these expenses.

"I'm not saying that expenses aren't important, but we haven't spent nearly enough time trying to estimate the revenue from our weekly offertory collections. In the past, we've basically just estimated whatever revenue we needed to cover the estimated expenses and plugged that number into the budget without any real thought as to whether we could actually expect to collect those amounts. We need to start the budget process with the revenue piece this year and make sure that we have a realistic estimate of our collections. Then we can try to estimate expenses that coincide with these revenue projections."

Fr. Jarreau liked what he had just heard from John and Charlie. It was obvious to him that the old budgeting process had some fatal flaws which could not be allowed to continue. The advisors' recommendations made a lot of sense to him. He knew that the department heads would complain about the significant spending cuts that would likely be required with a more realistic revenue estimate, but the financial viability of the entire parish was at stake. The department heads would simply have to prioritize between expenditures that they absolutely had to make and those that they could live without. It seemed like a much better idea to allow increased spending later on if collections turned out to be higher than expected, rather than have to cut expenditures that the department heads had planned to make as of the beginning of the year.

Budgeting for 2006

Because the budgeting process was scheduled to begin at the November meeting of the finance committee, Fr. Jarreau decided to call an additional meeting in the meantime to inform the committee about the new focus on revenue projection during the budgeting process. At this meeting, he asked the group for suggestions about how the offertory revenue could be predicted.

Frank Lawson, a vocal member of the committee but one who usually spent more time looking at his watch at meetings than actually contemplating the issues at hand before he spoke, characteristically blurted, "Why don't we just use the current year's actual offertory and be done with it? Whatever we collected this past January can be the forecast for this coming January. We should be spending more time thinking up additional ways to raise money beyond the collection basket. We need to be increasing the revenue to enable us to meet the expenses that we have now."

Trying in vain to conceal her exasperation, Megan Fisher, demand planner at a local consumer packaged goods company, responded quickly,

"You've been on this committee long enough, Frank, to know that the offertory collection is overwhelmingly the largest component of the church's total revenue. Any additional fundraising that we do is fine, but it's not going to totally make up for offertory projections that are way off.

"One of the most important parts of my job is to produce weekly forecasts of demand for our various product lines to make sure that we plan to have enough units to satisfy our customers. Why don't I take some time over the next few weeks and use some of the models that I use at work to forecast the offertory collections for each month of next year? That can be a starting point for our budget meeting in November."

Megan turned to Ernie Jackson, the church's bookkeeper. "Ernie, how much past data about the offertory collections can you get me? The more the better!"

"I think I can get you the last four years' worth of data. That shouldn't be a problem. Oh, and I'll also get you a list of the dates of the Holy Days for each year. That should have some kind of effect on the collections in those months because parishioners are obligated to attend Mass those days."

"Sounds good, Ernie. That's a great point about the Holy Days. I wasn't thinking about those. I wonder if the offertory revenue is related to any other factors. I'm going to have to think about those when I run the models. I'll let you know if I need any more data from you."

As the meeting wound down, Fr. Jarreau started to feel a little better about the church's future. Certainly some difficult financial decisions were on the horizon, but at least the committee had a plan that they were committed to and should help to stabilize the net cash outflows. He prayed that Megan would get the whole budgeting process off to a good start by producing a good forecast.

Exhibits

Exhibit 3.1 Monthly Offertory Revenue from July 2001 to September 2005

Month	2001	2002	2003	2004	2005
January		$110,492.56	$92,298.44	$98,005.33	$131,627.02
February		$90,979.03	$78.930.37	$114.943.12	$90,711.98
March		$128,952.91	$111,539.47	$88,289.13	$108,976.43
April		$79,301.47	$102,117.76	$100,502.85	$123,005.88
May		$76,936.52	$79.484.64	$111,646.53	$93,311.73
June		$94,806.21	$101,758.24	$83,580.73	$82,907.85
July	$99,061.10	$77,038.89	$85,851.77	$81,039.41	$97970.72
August	$89,066.57	$82,764.19	$98.602.05	$107,677.54	$78,723.84
September	$115,003.28	$104,756.91	$79,139.66	$85,619.97	$83,625.10
October	$86,224.72	$79,724.52	$79,178.51	$111,837.81	
November	$92,264.05	$96,470.47	$115,691.27	$82,599.90	
December	$181,938.85	$160,005.98	$155,950.77	$158,685.01	

Exhibit 3.2 Graph of Monthly Offertory Revenue from July 2001 to September 2005

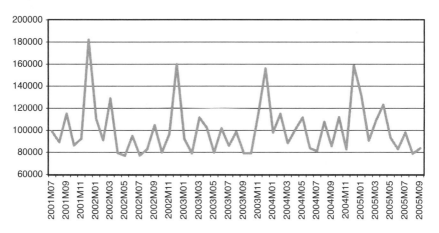

Exhibit 3.3 List of Catholic Holy Days of Obligation or Major Feast Days from 2001–2006

Holy Day/Feast	2001	2002	2003	2004	2005	2006
Solemnity of Mary	Jan 1	Jan 1	Jan 1	Jan 1	Jan 1	Jan 1
Ash Wednesday	Feb 28	Feb 13	Mar 5	Feb 25	Feb 9	Mar 1
Easter Sunday	Apr 15	Mar 31	Apr 20	Apr 11	Mar 27	Apr 16
Ascension	May 24	May 9	May 29	May 20	May 5	May 25
Assumption of Mary	Aug 15	Aug 15	Aug 15	Aug 15	Aug 15	Aug 15
All Saints' Day	Nov 1	Nov 1	Nov 1	Nov 1	Nov 1	Nov 1
Immaculate Conception	Dec 8	Dec 8	Dec 8	Dec 8	Dec 8	Dec 8
Christmas Day	Dec 25	Dec 25	Dec 25	Dec 25	Dec 25	Dec 25

Case 4

Pizza Station

Kathryn Marley, Duquesne University
Gopesh Anand, University of Illinois at Urbana–Champaign

Background

Established in 1980, Pizza Station is located in the trendy downtown area of Salina, Pennsylvania. Situated within walking distance of Salina State College, the restaurant initially offered in-house dining and a variety of food items on its menu. However, as competition among local restaurants grew, Pizza Station's staff decided to limit their offerings to delivery of pizzas in early 2001. Since then, they have developed a loyal following among customers who have come to expect quick and reliable delivery of good-quality pizza from the restaurant. Nevertheless, in the past two years, manager Tom Smith has noticed that customer complaints have increased significantly. With new pizza outlets and other restaurants opening up in the area every year, Tom is concerned that unless changes can be made quickly, Pizza Station will lose market share, and might eventually have to close its doors permanently.

Pizza Station operates seven days a week. On Sunday through Thursday, the hours are noon through 1 a.m. On Fridays and Saturdays, the hours are noon through 3 a.m. The busiest hours are on Fridays and Saturdays between 9 p.m. and 2 a.m. Demand for pizza varies throughout the week and times of day. From Sunday through Thursday, the average daily demand is 300 pizzas. Fridays and Saturdays are busier, with average demand increasing to 650 pizzas per day. On these two days, the demand during each busy 9 p.m.–2 a.m. period averages 400 pizzas. Currently, Pizza Station is promising a delivery time of 45 minutes to customers.

Tom recently hired Kate Fox, a business major from Salina State College, to manage the weekend shift. He asked her for assistance in identifying the necessary changes that would enable Pizza Station to decrease complaints, increase customer satisfaction, and win back lost customers. Kate recently completed a Lean Six Sigma training course as one of her business school classes; and, eager to apply some of the things she learned, she sat down with Tom to discuss the situation.

"I don't know where we went wrong and, frankly, I don't know where to begin!" exclaimed Tom. "All I know is that our troubles seem to have suddenly multiplied since last January when the students came back to campus. Things started off normally, but over the next three months I noticed a steadily increasing number of complaints." Tom pulled out a file folder from the bottom of a stack on his desk. Inside were papers of varying sizes with notes scribbled on them. He squinted as he tried to read them. "This customer said the crust was too thin, while this one said the delivery time was too long." As he read from the stack of mismatched notes, Kate realized that they were going to have to implement a better system for capturing customer feedback—and fast, if they were going to turn this place around.

"Okay, Tom, I get the idea," said Kate. "Let's start at the beginning. We need to approach this problem from a systematic process improvement perspective—which first involves figuring out what is the voice of the customer." Tom looked confused. Kate continued, "The voice of the customer (or VOC) consists of customer requirements, which is what the customer is expecting Pizza Station to deliver. There is no chance that customers are going to keep ordering pizza from Pizza Station if these expectations are not being met. So you need to capture this information to know where to begin to make changes."

"Sounds great, Kate," said Tom. "Let's get started!"

A customer satisfaction study was commissioned to figure out the voice of the customer (VOC). It pointed to delivery time and crust thickness as being critical to quality (CTQ) characteristics. An analysis of recent sales data revealed that the most commonly ordered crust from Pizza Station was the unique medium crust. In addition, three focus groups with eight customers each revealed that the ideal medium pizza crust was found to be between 4.25 mm and 5.75 mm. To measure what the process was actually producing (voice of the process, or VOP), Kate took a sample of five medium pizza crusts every day over a period of 30 days, and measured their thickness. The data that she collected is provided in Exhibit 4.1.

Kate's training in Lean Principles also prompted her to talk with the employees who actually work on the pizza-making line. As she told Tom, people working on the frontlines of any process know the most about how the work is done. From the spirited discussion that Kate had with the staff, it soon became apparent that they believed the task of order-taking had problems. So, she asked them to collect data on this task. For 30 days, they took samples of 50 orders every day, inspected them, and recorded all the errors involved. This data is provided in Exhibit 4.2.

Next, choosing one of the busiest times at Pizza Station, Kate walked the process to map the value stream for pizza-making and delivery. She explained to Tom that this exercise was aimed at 1) getting some measurements of different tasks in the process, and 2) gaining additional insights into the current length of delivery time. Because the peak demand period for Pizza Station is Friday nights between 9 p.m. and 2 a.m., Kate and Tom walked the process at that time and observed the following steps involved in making pizzas. Their observations are described in the following sections.

There are a total of five employees in the store on Friday nights, along with nine delivery drivers on staff. The pizza-making process begins with orders received by phone. Next, pizzas are assembled and baked. Finally, the pizzas are cut, boxed, labeled, and delivered to the customers.

Ordering

There are no designated operators who answer the phones at Pizza Station. The phones are answered by whoever is "nearby" at the time. That can be any of the four employees who are working the pizza line during the shift, with the exception of the employee who is dedicated to the baking process. It is estimated that each of the four employees spends 20% of her time answering phones. Kate watched the order-taking process for 15 minutes. During that time, orders for 20 pizzas were received. After the customer places the order, the employee who took the order informs the customer about the price and the estimated delivery time. The order is written on a note pad and hung on a board for the assembly station workers to retrieve as they become available. Kate observed during those 15 minutes that there were orders for 5 pizzas waiting to be assembled; an order waits on average 225 seconds before moving to assembly.

Pizza Assembly

Each of the four employees dedicates 60% of her time to assembling pizza. After an order is received, an assembly worker retrieves a ball of pizza dough from the refrigerator at the back of the store. Employees only retrieve one ball of pizza dough at a time. It takes, on average, three minutes for one employee to walk to the refrigerator and back every time an order is received. The worker begins by flattening the dough to the desired thickness and forming the crust. Next, the worker drizzles oil on the crust and assembles the pizza, which includes adding sauce, cheese, toppings, and seasonings. Kate noticed that this process took an average of 90 seconds per pizza. After the pizza is completed, it is placed on a tray until there is an available rack in the oven. Kate observed six pizzas waiting to be baked. A pizza waits 270 seconds before baking on average.

Baking

The oven used for baking the pizzas is set at 500 degrees to ensure crisp and efficient baking. The baking process takes nine minutes and the oven can hold six pizzas at a time. There is one worker at the oven station, who dedicates 100% of her time to putting the pizzas in the oven and removing them, as well as monitoring the baking time of each pizza. After the pizza is baked, it is removed and placed on a large wooden tray. Kate observed 15 pizzas waiting to be cut. A pizza waits an average of 675 seconds before moving to the next station.

Cutting/Boxing/Labeling

The four employees dedicate 20% of their time to cutting and boxing the pizzas. When the pizza is removed from the oven, a worker uses a metal cutter to cut the pizza into the appropriate number of slices. Then he assembles the box and places the cut pizza in the box. The employee then goes to the order station, retrieves the address information from the order slip, and writes this information on the pizza box. The boxed pizza is placed on the delivery counter and waits for delivery. Cutting the pizza takes an average of 5 seconds. Making the box takes 20 seconds. Placing the pizza in the box and closing the lid takes 5 seconds on average. Retrieving the order information and writing the information on the box takes 90 seconds on average. Kate observed 10 pizzas waiting to be delivered, and that, on average, the pizzas wait 450 seconds before being taken out for delivery.

Delivery

There are nine dedicated delivery drivers. When a driver returns from a delivery, she checks to see whether there is another pizza waiting to be delivered. If there is, she proceeds with delivering that pizza. If there is not, she is free to wait in the break room until there is a pizza waiting to be delivered. Each delivery driver delivers one order at a time. Because the majority of Pizza Station customers live within a five-mile radius of the restaurant, a worker is usually able to deliver a pizza and be back in the restaurant within 18 minutes; therefore, the delivery time to the customer is approximately 9 minutes.

Suppliers

Tom Smith is in charge of ordering all the supplies and ingredients for the restaurant. The pizza dough is made by a local bakery and delivered once a week by truck to Pizza Station. Each delivery consists of 2,500 pizza crusts. After the crusts are received, they are stored in a large refrigerator in the back of the storeroom. Because the delivery day varies, on average there are 1,250 balls of dough on hand.

Analysis

After walking the process with Tom, Kate started scrutinizing the current state of operations as depicted in the value stream map to consider potential areas of improvement. Tom was skeptical but excited.

"Let's get going, Kate; I don't want to lose one more customer if I can help it. Let's get Pizza Station back on top where we belong!"

Assume you are a new addition to Kate's Lean Six Sigma team. Answer the following questions:

1. Based on the sample data on pizza thickness collected for the 30-day period and presented in Exhibit 4.1, construct an X-bar–R chart. Include the information on customer requirements obtained through the focus groups, conduct a process capability analysis, and interpret its result.

2. Based on the data on errors in order-taking provided in Exhibit 4.2, construct a Pareto chart to identify the areas that should be top priorities for Pizza Station.

3. Compute the DPMO and sigma level of the order-taking task using the data in Exhibit 4.2.

4. Using the data on total daily errors provided in Exhibit 4.2, conduct an analysis of variance (ANOVA) test to determine whether there is a significant difference in errors on different days of the week.

5. Construct the appropriate control chart (based on the nature of the data collected) for the total number of defects or errors per 50 orders shown in Exhibit 4.2. Is this process in statistical control?

6. What is the Takt time for this process (in seconds)? (Note: Because the information was gathered during the Friday evening shift, use that time period for this analysis.)

7. Draw a Current State Map of this pizza-making process. Pizza Station is quoting a delivery lead time of 45 minutes to its customers. What is the total lead time between ordering and delivery? Is Pizza Station capable of meeting this promise based on your Current State Map calculations?

8. Develop a list of the symptoms that indicate problem areas in the pizza-making and delivery value stream. Provide suggestions for how the problems underlying the symptoms might be reduced or eliminated.

9. Draw a Future State Map of the pizza-making process, incorporating the changes you suggested in #8.

10. Develop an implementation plan for your suggested improvements at Pizza Station.

Exhibits

Exhibit 4.1 Pizza Crust Thickness (in Millimeters)

Sample Number	Observations				
	1	2	3	4	5
1	5.33	5.44	5.21	5.3	5.2
2	5.11	5.09	4.92	4.97	4.82
3	5.12	4.85	4.92	5.02	5.03
4	4.82	4.92	5.01	5.2	5.19
5	5.46	5.47	5.32	5.42	5.11
6	5.01	5.21	5.24	5.26	5.31
7	5.21	5.24	5.33	5.41	5.55
8	4.92	4.81	4.94	5.01	5.21
9	5.12	5.31	5.41	5.25	5.34
10	5.12	5.11	5.42	5.34	5.32
11	5.62	5.43	5.21	5.19	5.18
12	5.24	5.41	5.42	5.31	5.5
13	4.9	4.82	5.01	5.21	5.01
14	5.21	5.55	5.41	5.32	5.42
15	5.21	5.32	5.45	5.56	5.01
16	5.01	5.21	5.31	5.49	5.32
17	4.91	5.21	4.81	5.24	5.34
18	5.04	5.14	5.17	5.32	5.41

Sample Number	Observations				
	1	2	3	4	5
19	5.26	5.32	5.41	5.56	5.21
20	4.81	4.92	5.03	5.24	5.14
21	4.95	5.19	4.89	5.31	5.21
22	4.98	5.23	5.21	5.24	5.22
23	5.41	5.42	5.55	5.11	5.14
24	4.91	5.21	5.34	5.11	5.01
25	5.12	5.13	4.98	4.81	4.91
26	4.91	5.12	5.21	5.03	5.21
27	5.24	5.34	5.56	5.33	5.31
28	4.91	5.12	4.94	5.32	5.21
29	5.3	4.92	4.99	5.01	5.02
30	4.91	5.21	5.34	5.44	5.55

Exhibit 4.2 Errors in the Order-Taking Task

Sample Number	Number of Orders	Incomplete Address	Out of Range Address	Toppings Unclear	Missing Coupon	Forgot to Record Time	Others	Total	Days of Week
1	50	1	7	0	1	5	0	14	Monday
2	50	0	5	0	0	1	1	7	Tuesday
3	50	0	0	2	2	3	1	8	Wednesday
4	50	1	6	1	0	8	0	16	Thursday
5	50	2	14	1	0	18	3	38	Friday
6	50	0	1	0	2	8	4	15	Saturday
7	50	0	1	1	1	3	0	6	Sunday
8	50	2	6	1	1	11	0	21	Monday
9	50	0	7	1	0	2	1	11	Tuesday
10	50	0	3	0	0	3	0	6	Wednesday
11	50	1	7	0	1	1	0	10	Thursday
12	50	1	18	1	0	11	0	31	Friday
13	50	2	9	2	0	0	0	13	Saturday
14	50	0	3	0	0	2	3	8	Sunday
15	50	0	5	2	0	6	1	14	Monday
16	50	3	2	1	1	3	0	10	Tuesday
17	50	2	3	2	1	9	1	18	Wednesday
18	50	1	8	0	0	5	3	17	Thursday

Case 4 • Pizza Station

Sample Number	Number of Orders	Incomplete Address	Out of Range Address	Toppings Unclear	Missing Coupon	Forgot to Record Time	Others	Total	Days of Week
19	50	1	18	1	0	3	1	24	Friday
20	50	0	1	0	1	2	0	4	Saturday
21	50	0	1	2	1	8	3	15	Sunday
22	50	0	7	1	1	9	2	20	Monday
23	50	0	0	2	0	1	0	3	Tuesday
24	50	1	1	2	2	2	0	8	Wednesday
25	50	2	9	1	1	6	1	20	Thursday
26	50	1	12	0	0	12	2	27	Friday
27	50	0	8	1	0	5	1	15	Saturday
28	50	0	10	0	0	4	1	15	Sunday
29	50	0	4	2	0	9	2	17	Monday
30	50	0	4	1	0	8	1	14	Tuesday
Totals		21	180	28	16	168	32	445	

2
Optimization and Simulation

5. Inventory Management at Squirrel Hill Cosmetics — 47
6. Safety Stock Planning for a Hong Kong Fashion Retailer — 65
7. Network Design at Commonwealth Pipeline Company — 77
8. Publish or Perish: Scheduling Challenges in the Publishing Industry — 81

Case 5

Inventory Management at Squirrel Hill Cosmetics

Paul M. Griffin, Pennsylvania State University

Company Background

Founded in 1978, Squirrel Hill Cosmetics, Inc., is a privately held company that sells cosmetic products including eyeliner, mascara, lipstick, face powder, and nail polish. The company originally made its name with a popular medicated lip balm that it manufactured in Pittsburgh. In the 1980s and 1990s the company grew primarily through acquisitions and rapidly increased its product offerings. By 2003, most of its production was moved to China, and currently it contracts out all of its manufacturing.

Squirrel Hill Cosmetics now manages more than 1,200 SKUs; approximately 15% of the SKUs turn per year because of changing customer preferences. Last year the company made roughly $200 million in profit with approximately 900 employees. The two channels that it sells through are retailers, with Walmart and Target being the two key customers, and drugstores, including CVS and Walgreens.

In 2010, Squirrel Hill Cosmetics significantly redesigned its supply chain. More than 95% of its SKUs are sourced from China, and these products are now brought by container to a large central distribution center (DC) in Nanty-Glo, Pennsylvania. At the DC, some of the products are packaged for specific customers, such as CVS. Products are then typically shipped to the regional DCs of its customers, although this does vary some by customer.

The containers brought to Nanty-Glo from China have approximately a 12-week lead time. The average cost per container is $6,800, and approximately 1,000 containers were shipped this past year. For key customers, namely Walmart, Nanty-Glo will emergency ship product by air. This reduces the 12-week lead time to 2 days, assuming that the Chinese supplier has available stock. Last year, expediting orders through air freight was done for roughly 20% of the backordered items. Shipping products by air costs 56.3% more. The average value of an item across all SKUs is $1.

Shortly after the supply chain redesign, Squirrel Hill Cosmetics updated its business management software to SAP. Although there were some significant hiccups in the implementation, the software was running smoothly in about six months. Squirrel Hill is now much better at tracking information and integrating it with its financial and operational systems.

Current Operations

Kim Deal was recently promoted to the position of Vice President of Global Supply Chain Operations. The promotion was due in large part to her successful modernization efforts of the Nanty-Glo DC. This included the design and installation of a large sortation process and supporting software, which will meet its payback period in about half the company standard of two years. She has also built a solid reputation as someone who is quite good at accurately estimating the financial impact of engineering changes.

Within two weeks of taking her new position, Kim felt that the biggest opportunity was to better manage global inventory and the related distribution costs, particularly for those items sourced from

China. To better understand the inventory operations, Kim set up a meeting with Tanya Donelly, inventory manager of the Nanty-Glo DC. Kim knew Tanya fairly well, as they had worked together on several projects in the past, although Kim had not directly worked on inventory-related projects.

During the meeting, Tanya explained the basic inventory management process currently in use. In particular, she walked through a specific example for SKU QED0001 (provided in the "Appendix: Inventory Example Given by Tanya Donelly" at the end of this case). Tanya pointed out that the beauty of the system is its simplicity; the same approach can be used for all the SKUs, and all that is really required is a good forecast.

Kim had several questions about the system. In particular, she was interested in why a four-week interval was chosen for the desired safety stock level for a SKU in each period. Tanya replied that this was something that they experimented with when they first implemented SAP, and they felt that a four-week interval gave the best performance results.

The financial performance of the system was another area about which Kim had several questions. In particular, she wanted to know the annual inventory holding cost, the backorder cost, the distribution cost, and the expediting cost for items that are backordered to key customers. Tanya replied that over the past year, there were 934 containers that had been shipped from China to the United States, resulting in a total shipping cost of $6,351,200. Further, there was a budget of $400,000 allocated for air freight of expedited orders for the past year, but the actual total was roughly $1.1 million (and roughly 14.73% of all items were shipped by air freight). Tanya also mentioned that they do not track holding or backorder costs, but only units; with the relatively new SAP system, they do not rely on such cost estimates to operate. She also pointed out that in her view, this was one of the benefits of their system; namely, only units need to be tracked, and costs do not need to be estimated.

Potential Problems

Kim left the meeting with Tanya somewhat troubled by what she had heard. Kim was particularly bothered by the lack of financial performance measures. One of her first actions was to determine how good the forecasts were for their China-sourced products. Kim pulled two years of weekly data on 200 different SKUs and found that there was unfortunately not much correlation between forecasted and actual demand. As the entire inventory model relied on having an accurate forecast, this left Kim with two options for what to do next: work on improving the forecasting system or develop a new way to manage the inventory. Both of these were important to her; however, she realized that even if she had a perfect forecasting system, she still did not feel comfortable with the current inventory management system.

Three aspects of the current system bothered her. First, all of the SKUs are treated the same, regardless of their demand characteristics. Second, if certain performance characteristics such as inventory holding and backordering are not measured financially, how can appropriate tradeoffs be considered? It seemed clear to her, for example, that holding additional inventory for a few items might greatly help to reduce the expediting costs from backordering. Finally, since they were not particularly good at forecasting, could they design a system that was robust to the forecast error?

Kim decided that the inventory management system needed to fundamentally change. She wanted it to be driven by financial tradeoffs and yet be simple enough that it could be easily implemented within the SAP system. She had already had enough bad experiences with projects that led to costly consulting fees to modify the SAP implementation. She contacted Tanya to set up a meeting to discuss her ideas.

A Second Meeting

Kim explained to Tanya her concerns about the current system. Although Tanya listened politely, she responded that they were coming into the busiest time of the year and that she really did not have the time to devote to making any changes. In addition, she explained that the systems had worked well for more than a year and that there were certainly other better areas on which Kim could spend her time, such as renegotiating the contracts with their Chinese shipping container lines.

This response did not sit well with Kim. She stated rather firmly that overspending their budgeted expedited shipments by more than $600,000 should be sufficient cause for alarm to anyone about the lack of effectiveness of the current inventory management practices. She also informed Tanya that she was making this project a priority. Although she understood that Tanya has tremendous pressures in Nanty-Glo, she stated that she has tremendous respect for her abilities and would like her to consider taking on the project. The expectation was that Tanya would agree.

The Memo

After the second meeting, Kim sent a memo to Tanya detailing her desired actions.

TO: Tanya Donelly, Inventory Manager–Nanty-Glo, PA

FROM: Kim Deal, VP of Global Supply Chain Operations

DATE: July 8, 2013

RE: Inventory Management Process

As I stated in our meeting on July 1, I believe we need to fundamentally change our inventory management system. I firmly believe that significant savings *and* improved performance can be achieved if done well. I am therefore putting you in charge of a project to develop it. I will make whatever resources you need available to you, but I want this to be your top priority. In particular, I would like you to take the following actions:

- Develop an appropriate way for us to estimate the holding, transaction (or fixed), and backorder costs. I want holding and backorder costs to be expressed at the unit level.
- Customer service is important to us, but it comes at a cost. Discuss the relationship between the service level and shortage cost (fractional charge per unit short). Discuss how we might establish this relationship quantitatively.
- I have been reading quite a bit about continuous and periodic review order systems, and believe this would be appropriate for us to use. Develop a system for us that uses a continuous review approach. I am particularly interested in making shortage costs a part of this. I also want to be able to see what the impact of setting different service levels would be. As you know, I understand things better by example. Please illustrate the method for the QED0001 example that you showed me using the following data (assume annual demand is 484,119):

Week	1	2	3	4	5	6	7	8	9	10	11
D	10,124	15,432	13,988	9,845	6,756	5,322	4,988	5,796	7,013	5,932	5,844
Q	19,697	15,004	10,951	6,685	5,445	4,722	5,186	5,773	7,950	7,363	6,492

- For this example, determine the probability of a stockout and an average holding cost (on hand plus on order). Further, use this example to discuss sensitivity of the holding, fixed, and backorder costs. Discuss how this might compare to our current system.

- I have done some preliminary work and noticed that many of our SKUs have quite different values for variability of demand over our lead time. Looking at the data, I picked three SKUs that I believe are representative and have given you their historical data in an attached appendix. For this example, please determine the current holding cost and emergency ship (expediting) cost. To easily compare with our current method, apply the periodic review method (period = 1 week) to the data and compute the same estimates. Discuss the differences. Assume that all backordered items use expedited shipping. Note that the analysis should be done for the data over weeks 1 to 52. I have given you the previous 12 weeks of order quantities for this year and the forecast for the 12 weeks following this year to enable you to compute all the estimates for the year of interest (weeks 1 to 52). Assume that an item of each SKU has a value of $1 to us.

- Give me your overall recommendations and next steps based on this preliminary analysis. Make sure to address how this new approach would be implemented into our SAP system.

- I realize that I am asking you to focus fairly narrowly on inventory here. I am also interested in your thoughts about other areas that you believe could lead to savings.

Please be prepared to give a 20-minute presentation of your findings at the next operations review meeting in two weeks. In addition, please prepare a brief report to me that addresses each of the preceding points. The report should be no more than six pages in length. Please be clear about any assumptions you made in your analysis.

I appreciate your effort with this, Tanya. Please let me know at your earliest convenience whether you will accept this assignment.

Appendix to the Memo

ITEM: SKU1							
Week	Forecast	Desired Safety Stock	Demand	Order Quantity	Receipts	Projected On Hand	On Hand Inventory
-12				7200			
-11				7200			
-10				7200			
-9				7200			
-8				7200			
-7				7998			
-6				8013			
-5				8900			
-4				12063			
-3				14229			
-2				14388			
-1			5804	12088	5905		24863
1	5869		6034			28800	
2	6802		5788			29598	
3	6905		8834			30411	
4	5884		15867			32111	
5	6994		8834			36974	
6	7998		8664			43205	
7	8013		8011			49580	
8	8883		7600			52768	
9	12063		14385			53230	
10	14986		16788			44055	
11	15086		23855			33158	
12	13092		19063			27893	
13	8906		7883				
14	8834		8850				
15	7867		7900				
16	7554		7600				
17	5432		6012				
18	4889		4800				
19	4505		4833				
20	4894		5300				
21	7865		7924				
22	8988		8895				
23	9324		9425				
24	9959		10011				
25	8834		9233				
26	7986		10000				
27	5986		6102				
28	7800		7764				
29	7993		7783				
30	7885		7924				

Case 5 • Inventory Management at Squirrel Hill Cosmetics

ITEM: SKU1							
Week	Forecast	Desired Safety Stock	Demand	Order Quantity	Receipts	Projected On Hand	On Hand Inventory
31	6799		8233				
32	8900		9683				
33	9974		9734				
34	10111		12228				
35	17899		19362				
36	9088		10001				
37	5788		6703				
38	5988		6802				
39	10888		11234				
40	11645		9854				
41	17893		15987				
42	14867		10124				
43	17683		15432				
44	14888		13988				
45	10997		9845				
46	6895		6756				
47	5406		5322				
48	5509		4988				
49	5586		5796				
50	6013		7013				
51	5833		5932				
52	5900		5844				
53	5579						
54	5777						
55	6830						
56	5607						
57	6994						
58	7998						
59	8013						
60	8883						
61	12063						
62	16333						
63	16092						
64	13009						
65	14897						

ITEM: SKU2							
Week	Forecast	Desired Safety Stock	Demand	Order Quantity	Receipts	Projected On Hand	On Hand Inventory
-12				21000			
-11				19000			
-10				20000			
-9				22400			
-8				23000			
-7				18500			
-6				18500			
-5				23200			
-4				22000			
-3				19800			
-2				18900			
-1			7903	19500	19800		48900
1	18048		13596			55859	
2	16544		26040			52266	
3	12147		17349			55203	
4	15332		26569			48583	
5	18867		29171			43020	
6	10915		19227			57940	
7	18463		31693			45650	
8	28714		26719			59508	
9	23079		19571			55546	
10	26392		25205			40862	
11	20492		17656			42945	
12	17281		25411			47228	
13	19916		28026				
14	22518		23522				
15	16028		25956				
16	18210		30928				
17	24475		24065				
18	29684		20359				
19	14964		32446				
20	24153		32273				
21	23471		31112				
22	18651		17215				
23	14531		32469				
24	17167		27982				
25	24677		32131				
26	21127		29969				
27	15457		15231				
28	21385		25198				
29	18136		31840				
30	17581		24794				

CASE 5 • INVENTORY MANAGEMENT AT SQUIRREL HILL COSMETICS

ITEM: SKU2							
Week	Forecast	Desired Safety Stock	Demand	Order Quantity	Receipts	Projected On Hand	On Hand Inventory
31	17810		31123				
32	19237		16860				
33	15357		28592				
34	21852		19442				
35	29990		19763				
36	27716		13088				
37	10827		18446				
38	23051		21765				
39	29715		32187				
40	28215		30316				
41	12909		19273				
42	11684		29643				
43	19830		28751				
44	15391		23077				
45	28712		21594				
46	18455		22422				
47	11844		23479				
48	17652		20201				
49	20804		21411				
50	18589		17494				
51	11388		20836				
52	25689		16501				
53	11469						
54	13588						
55	11160						
56	27316						
57	22735						
58	10744						
59	14094						
60	12357						
61	15138						
62	26594						
63	18250						
64	25568						
65	25938						

ITEM: SKU3

Week	Forecast	Desired Safety Stock	Demand	Order Quantity	Receipts	Projected On Hand	On Hand Inventory
-12				10100			
-11				7200			
-10				11000			
-9				9000			
-8				11000			
-7				10050			
-6				9020			
-5				8900			
-4				12063			
-3				10000			
-2				7200			
-1			7903	9500	9800		24863
1	11032		8949			38200	
2	9754		12036			41050	
3	10743		7935			39070	
4	11141		7635			38970	
5	11772		12242			40033	
6	9237		11616			39983	
7	10650		8395			38163	
8	8276		11867			38763	
9	11200		9448			60158	
10	11033		8011			57358	
11	10662		12062			61746	
12	9698		10943			59495	
13	8143		10630				
14	11995		10138				
15	9342		9917				
16	9422		11321				
17	11151		11285				
18	8721		8666				
19	9868		11035				
20	10057		9935				
21	10822		12066				
22	9703		10529				
23	8203		9278				
24	10521		10731				
25	10315		9605				
26	9834		8276				
27	8171		10223				
28	8494		7670				
29	10535		11950				
30	8633		8846				

CASE 5 • INVENTORY MANAGEMENT AT SQUIRREL HILL COSMETICS

ITEM: SKU3							
Week	Forecast	Desired Safety Stock	Demand	Order Quantity	Receipts	Projected On Hand	On Hand Inventory
31	10301		10849				
32	8661		9665				
33	8422		8326				
34	9104		7909				
35	8768		11627				
36	11600		8980				
37	10779		10422				
38	9944		9770				
39	11759		10990				
40	9646		7718				
41	11108		8817				
42	8616		12372				
43	9013		11047				
44	10326		9416				
45	10446		8959				
46	9442		9667				
47	10010		12218				
48	11198		8755				
49	11428		7963				
50	11512		9787				
51	11733		9892				
52	8424		11971				
53	10912						
54	9771						
55	9461						
56	8230						
57	11753						
58	11769						
59	11997						
60	8488						
61	11689						
62	9315						
63	8364						
64	9208						
65	11261						

Appendix: Inventory Example Given by Tanya Donelly

Squirrel Hill Cosmetics manages more than 1,200 SKUs that are sourced from China with a lead time of 12 weeks. For each SKU, it holds 28 days (4 weeks) of safety stock. Its ordering policy is set up to maintain this desired safety stock level. A minimum order quantity of 7,200 units is used.

An example for SKU QED0001 is presented to illustrate the policy. It has a unit cost of $1.53, and makes up roughly 1.5% of total sales. The notation used in the example is as follows:

L = Lead time (in weeks; it is equal to 12)

F_t = Forecasted demand for period t (in units)

DSS_t = Desired safety stock level for period t (in units)

OH_t = On-hand inventory level in period t (in units)

POH_t = Projected on-hand inventory level in period t (in units)

Q_t = Order quantity in period t (in units; this is the decision variable)

The formulas used to compute the order quantity are (note that 12 is used for L) as follows:

$$Q_t = F_{t+12} + (DSS_{t+12} - POH_{t+12})$$

$$DSS_{t+12} = \sum_{i=t+12}^{t+12+4} F_i$$

$$POH_{t+12} = OH_t + \sum_{i=t-12-1}^{t-1} Q_i - \sum_{i=t}^{t+12-1} F_i$$

A completed spreadsheet for QED0001 is given in the following for a 26-week time period:

CASE 5 • INVENTORY MANAGEMENT AT SQUIRREL HILL COSMETICS

QED0001									
Week	Forecast	Desired Safety Stock	Demand	Order Quantity	Receipts	Projected On Hand	On Hand Inventory	Backorder	Expedite Quantity
1	5869	26585	6034	12525	7200	27824	24964	0	0
2	6802	27781	5788	5054	7200	26093	26130	0	0
3	6905	28889	8834	3491	7200	27705	27542	0	0
4	5884	31888	9267	6823	7200	32111	25908	0	0
5	6994	36957	8834	11248	7200	35099	23841	0	0
6	7998	43945	8664	10828	7998	41003	22207	0	0
7	8013	51018	8011	9990	8013	48067	21541	0	0
8	8883	55227	8900	9957	8900	50289	21543	0	0
9	12063	52070	14385	8851	12063	51077	21543	0	0
10	14986	45918	16788	10308	14229	45022	19221	0	0
11	15086	38699	23855	7788	14388	37044	16662	0	0
12	13092	33161	19063	16569	12088	27893	7195	0	0
13	8906								
14	8834								
15	7867								
16	7554								
17	5432								
18	4889								
19	4505								
20	4894								
21	7865								
22	8988								
23	9324								
24	9959								
25	8834								
26	7986								

In this example, all of the calculations are given for the first 12 weeks. Also, the order quantity arrives 12 periods later in the Receipts column. In addition, the forecast for the next 14 weeks is provided. The desired safety stocks and projected on-hand inventory levels can be computed for the remaining periods. We compute the desired safety stock for period 13 and projected on-hand inventory for period 25. The desired safety stock is based on the forecast.

For period 13, this would be computed by looking at the forecast over the next four weeks:

QED0001								
Week	Forecast	Desired Safety Stock	Demand	Order Quantity	Receipts	Projected On Hand	On Hand Inventory	Backorder	Expedite Quantity
1	5869	26585	6034	12525	7200	27824	24964	0	0
2	6802	27781	5788	5054	7200	26093	26130	0	0
3	6905	28889	8834	3491	7200	27705	27542	0	0
4	5884	31888	9267	6823	7200	32111	25908	0	0
5	6994	36957	8834	11248	7200	35099	23841	0	0
6	7998	43945	8664	10828	7998	41003	22207	0	0
7	8013	51018	8011	9990	8013	48067	21541	0	0
8	8883	55227	8900	9957	8900	50289	21543	0	0
9	12063	52070	14385	8851	12063	51077	21543	0	0
10	14986	45918	16788	10308	14229	45022	19221	0	0
11	15086	38699	23855	7788	14388	37044	16662	0	0
12	13092	33161	19063	16569	12088	27893	7195	0	0
13	8906	29687	=8834+7867+7554+5432		3		220	0	0
14	8834	25742	8850	6862	5054	29522	4862	0	0
15	7867	22380	7900	6815	3491	26756	1066	0	0
16	7554	19720	7600	7433	6823	20451	-1843	-1843	1500
17	5432	22153	6012	10020	11248	16337	-2620	-2620	0
18	4889	26252	4800	10691	10828	20313	2616	0	0
19	4505	31071	4833	17810	9990	25586	8644	0	0
20	4894	36136	5300	9416	9957	31073	13801	0	0
21	7865	37105	7924	6194	8851	36119	18458	0	0
22	8988	36103	8895	6047	10308	34783	19385	0	0
23	9324	32765	9425	10795	7788	34301	20798	0	0
24	9959	30606	10011	11746	16569	23996	19161	0	0
25	8834	29765	9233	17945	13964	24635	25719	0	0
26	7986	29664	10000	15266	6862	30788	30450	0	0

Given the forecast, all the desired safety stocks can be computed. Next, the calculation for projected on-hand inventory is shown for period 25. Note that for projected order quantities in earlier periods, the order quantities in earlier time periods given in the spreadsheet would be used. However, the calculation is the same.

CASE 5 • INVENTORY MANAGEMENT AT SQUIRREL HILL COSMETICS

QED0001

Week	Forecast	Desired Safety Stock	Demand	Order Quantity	Receipts	Projected On Hand	On Hand Inventory	Backorder	Expedite Quantity
1	5869	26585	6034	12525	7200	27824	24964	0	0
2	6802	27781	5788	5054	7200	26093	26130	0	0
3	6905	28889	8834	3491	7200	27705	27542	0	0
4	5884	31888	9267	6823	7200	32111	25908	0	0
5	6994	36957	8834	11248	7200	35099	23841	0	0
6	7998	43945	866 B	10828	7998	41003	22207	0	0
7	8013	51018	8011	9990	8013	48067	21541	0	0
8	8883	55227	8900	9957	8900	50289	21543	0	0
9	12063	52070	14385	8851	12063	51077	21543	0	0
10	14986	45918	16788	10308	14229	45022	19221	0	0
11	15086	38699	23855	7788	14388	37044	16662	0	0
12	13092	33161	19063	16569	12088	27893	7195	0	0
13	8906	29687	7883	13964	12525	26068	220	0	0
14	8834	25742	8850	6862	5054	29522	4862	0	0
15	7867	22380	7900	6815	3491	26756	1066	0	0
16	7554	19720	7600	7433	6823	20451	-1843	-1843	1500
17	5432	22153	6012	10020	11248	16337	-2620	-2620	0
18 A	4889	26252	4800	10691	10828	20313	2616	0	0
19	4505	31071	4833	17810	9990	25586	8644	0	0
20	4894	36136	5300	9416	9957	31073	13801	0	0
21	7865	37105	7924	6194	8851	36119	18458	0	0
22	8988	36103	8895	6047	10308	34783	19385	0	0
23	9324	32765	9425	10795	7788	34301	20798	0	0
24	9959	30606	10011	11746	16569	23996	19161	0	0
25	8834	29765	9233	17945	13964	24635	=7195+B-A		0
26	7986	29664	10000	15266	6862	30788	30450	0	0

The remaining calculations are the On-Hand Inventory (OH_t), Backorder Quantity (B_t), and Expedite Quantity (E_t). Expedite quantity is the amount backordered that is required at key customers. The remaining formulas are as follows, where receipts in period t are denoted R_t:

$$OH_t = OH_{t-1} + R_t - D_t$$
$$B_t = \min(0, OH_t)$$

The completed spreadsheet for the entire period is then as follows:

QED0001									
Week	Forecast	Desired Safety Stock	Demand	Order Quantity	Receipts	Projected On Hand	On Hand Inventory	Backorder	Expedite Quantity
1	5869	26585	6034	12525	7200	27824	24964	0	0
2	6802	27781	5788	5054	7200	26093	26130	0	0
3	6905	28889	8834	3491	7200	27705	27542	0	0
4	5884	31888	9267	6823	7200	32111	25908	0	0
5	6994	36957	8834	11248	7200	35099	23841	0	0
6	7998	43945	8664	10828	7998	41003	22207	0	0
7	8013	51018	8011	9990	8013	48067	21541	0	0
8	8883	55227	8900	9957	8900	50289	21543	0	0
9	12063	52070	14385	8851	12063	51077	21543	0	0
10	14986	45918	16788	10308	14229	45022	19221	0	0
11	15086	38699	23855	7788	14388	37044	16662	0	0
12	13092	33161	19063	16569	12088	27893	7195	0	0
13	8906	29687	7883	13964	12525	26068	220	0	0
14	8834	25742	8850	6862	5054	29522	4862	0	0
15	7867	22380	7900	6815	3491	26756	1066	0	0
16	7554	19720	7600	7433	6823	20451	-1843	-1843	1500
17	5432	22153	6012	10020	11248	16337	-2620	-2620	0
18	4889	26252	4800	10691	10828	20313	2616	0	0
19	4505	31071	4833	17810	9990	25586	8644	0	0
20	4894	36136	5300	9416	9957	31073	13801	0	0
21	7865	37105	7924	6194	8851	36119	18458	0	0
22	8988	36103	8895	6047	10308	34783	19385	0	0
23	9324	32765	9425	10795	7788	34301	20798	0	0
24	9959	30606	10011	11746	16569	23996	19161	0	0
25	8834	29765	9233	17945	13964	24635	25719	0	0
26	7986	29664	10000	15266	6862	30788	30450	0	0

Case 6

Safety Stock Planning for a Hong Kong Fashion Retailer

Tsan-Ming (Jason) Choi, The Hong Kong Polytechnic University[1]

Introduction and Company Background

Inventory management is a critical part of fashion retail supply chain management. For most fashion products, consumer demand is very difficult to forecast, which makes inventory planning more difficult. In this case analysis, we explore the safety stock management problem in a fashion retailer in Hong Kong. With the (adapted) real data from the company, a systematic analysis on safety stock level can be conducted.

JTMC[2] is one of the well-known chain store fashion retailers in Hong Kong. JTMC currently has 20 retail shops in Hong Kong[3] (each of which carries approximately 400 items), and its annual sales turnover is approximately U.S. $100 million. In addition, it also has an overseas retail network extending to China, Macau, Indonesia, Australia, and many countries in the Middle East, with

---

[1] Tsan-Ming (Jason) Choi can be contacted at jason.choi@polyu.edu.hk.
[2] Company JTMC is a fictitious name for a real company.
[3] To keep the company details anonymous, we modified and/or scaled the numbers provided in this case from the real values, but they do show the essential feature of the company and its inventory practice.

a combination of direct operation, franchising, and licensing. The company's objective is to provide its customers with high-quality service and products at a very reasonable price. One of the company's operational philosophies in terms of inventory management is that stock-outs should be avoided as much as possible. To achieve this target, the senior management of the company implements a logistics planning scheme in which (1) the product is replenished every day in a one-to-one replenishment manner (that is, one unit sold today will be filled by a replenishment within one day); and (2) inventory levels should be high and appropriately fill the available display space throughout the year for each in-season item. Thus, for both peak and non-peak seasons, the inventory level (including safety stock) for the same in-season item is the same (if the product is not out of stock, which is also the usual case for JTMC).

Safety Stock Analysis

We have collected a sample of demand data from JTMC (Exhibits 6.1 and 6.2) and inventory data (Exhibit 6.3) for 10 items. With the use of these data sets, we can study quantitatively the appropriate safety stock level with respect to a target inventory service level. We can also comment on the efficiency of the existing inventory planning practice in JTMC.

To conduct the analysis, we make a few assumptions. For the sake of simplicity, we assume the daily demand of each item during each selling season follows a normal distribution. Using the information from Exhibits 6.1 and 6.2, we can construct Exhibits 6.4 and 6.5, which show the mean and the standard deviation of the daily demand of each item directly.

Because replenishment lead time is fixed (in fact, it is equal to 1), we can employ the standard textbook safety stock formula to calculate the amount of safety stock needed with a given inventory service target. The specific formula is given here:

$$SS_i^k = \sigma_i^k \sqrt{L} \Phi^{-1}(\alpha)$$

where

- SS_i^k = the amount of required safety stock for item $i = 1, 2,..., 10$, during season $k = peak, non\text{-}peak$.
- σ_i^k = the standard deviation of daily demand for item $i = 1, 2,..., 10$, during season $k = peak, non\text{-}peak$.
- L = the lead time measured in days.
- α = the target inventory service level, which represents the chance of no stock-out during the replenishment cycle (that is, lead time L) and $\alpha < 100\%$.
- $\Phi^{-1}(\cdot)$ = the inverse of the standard normal cumulative distribution function (cdf).[4]

With (1), we can calculate the amounts of required safety stock for each item with different inventory service targets. The results are summarized in Exhibits 6.6 and 6.7.

With the amount of calculated safety stock as shown in Exhibits 6.6 and 6.7, we can establish the inventory level that should be kept for each item with a given inventory service target for each season scenario. In this case study, the inventory level is equal to the mean of daily demand multiplied by the lead time plus the amount of required safety stock, as shown in the following formula:

Inventory level of item i = Mean daily demand of item $i \times L$ + the amount of required safety stock of item i.   (2)

Exhibit 6.8 shows the result.

---

[4] In Excel, the inverse of the standard normal cdf can be computed by the built-in function "normsinv()." In this case, the numerical analysis is conducted by the Office 2010 version of Microsoft Excel. A different version of Excel might yield slightly different numerical values.

Define: The inventory saving (in quantity) = current inventory level in the company (Exhibit 6.3) − calculated inventory level with a given inventory service target (Exhibit 6.8).

From Exhibits 6.3 and 6.8, we can compute Exhibit 6.9, which summarizes the inventory saving in quantity.

It is obvious from Exhibit 6.9 that the company has overstocked the majority of the items under consideration; therefore, the current safety stock levels are not correctly determined. As a result, substantial inventory savings potentially exist even for the case when the target inventory service level is set to be very high. Notice that the only exceptions, which refer to the cases in which the company has understocked, appear in two cases (both highlighted in Exhibit 6.9: Items 1 and 2 under the peak season scenario). Thus, the current inventory management practice in the company is far from efficient. There is a big opportunity to improve, for example, by re-benchmarking the inventory level with respect to the theoretical benchmark in terms of the required inventory to achieve the specific inventory service target.

Inventory Cost Analysis

As shown in Exhibit 6.9, despite having significant inventory savings, JTMC senior management is also interested in estimating the monetary value associated with the potential inventory savings. If the monetary cost-savings realized by a change in inventory practice is small, the company might not have an incentive to implement the change. In the following section, we discuss how the company can estimate the monetary saving by changing the safety stock level with the theoretical benchmark. Suppose that according to the accounting manager of the company, the inventory holding cost of an item per season is estimated to be 2.8% of its product value. For the sake of simplicity, suppose that the product value for items 1 to 5 is $10 each, and for items 6 to 10, is $15 each. Exhibit 6.10 shows the monetary value of the inventory saving by adopting the theoretical benchmark for the 10 selected items under consideration in the study.

From Exhibit 6.10, we have two findings:

1. The monetary savings are larger when the target inventory service is smaller. This point can be explained by the fact that the company's current inventory level tends to be excessive.

2. Comparing between the peak season and the non-peak season, the monetary value of inventory savings is larger under the non-peak season case. This result is intuitive because the company's current practice simply sets the same inventory level regardless of whether the seasonal demand is high (peak season) or low (non-peak season). Despite being an easy measure, it is in fact very inefficient.

Further Discussions and Analysis

In the preceding sections, we investigated the case using the simple analytical safety stock formula with the given data. Some areas for further analysis are listed in the following:

1. In the preceding analysis, the estimated safety stock levels are not expressed in integer values. Suppose that inventory levels must be expressed as integers, and re-analyze the case assuming that the given target inventory service level is the "minimum" required target inventory service level.

2. Assuming that the 10 items are good representatives of the whole retail assortment of the company (in terms of inventory savings) and that the peak season occurs over four months of each year,[5] what will be the estimated annual monetary inventory savings for JTMC if it changes its safety stock quantity from the current practice to the theoretical benchmark if the target inventory service level is 95%?

---

[5] Notice that each season is one month long. Thus, the remaining eight months are non-peak seasons.

3. Discuss whether the one-day replenishment cycle under the current "everyday replenishment" inventory practice is necessarily the most efficient method.
4. If there is a certain level of uncertainty with respect to the lead time, how will this lead time uncertainty affect the inventory cost savings for JTMC if it changes its safety stock quantity from the current practice to the theoretical benchmark for a certain given inventory service level?

Exhibits

Exhibit 6.1 A Sample of Demand Data During the Peak Season for 10 Items of the Same Product Category (Adapted from and Scaled in Units with Respect to the Real Data Set)

Day	Item									
	1	2	3	4	5	6	7	8	9	10
1	1	2	2	0	1	1	3	3	2	2
2	1	3	1	1	2	1	1	4	2	1
3	0	1	3	2	1	1	2	2	1	3
4	2	1	1	1	0	2	2	3	4	2
5	2	4	3	2	3	1	3	4	2	0
6	3	2	2	1	1	0	4	4	3	2
7	2	1	2	0	0	3	5	2	5	2
8	2	2	1	1	2	0	3	4	3	0
9	3	1	0	3	1	4	5	6	4	3
10	1	0	1	1	3	0	5	2	3	4
11	0	2	1	2	0	0	6	1	6	1
12	3	1	3	1	1	2	5	2	4	0
13	1	2	2	2	2	4	4	4	4	3
14	4	1	0	1	0	3	3	5	3	5
15	1	3	3	2	0	0	5	6	4	2
16	3	2	1	0	3	5	2	3	3	5
17	1	1	0	1	1	2	3	3	6	0

Day	1	2	3	4	5	6	7	8	9	10
18	1	3	2	2	1	0	1	4	4	2
19	4	2	4	1	1	3	1	2	3	3
20	1	4	4	3	0	2	5	1	6	2
21	2	0	1	1	2	3	2	0	4	0
22	1	5	2	4	1	0	1	2	4	1
23	2	2	0	1	3	4	5	3	6	2
24	2	4	1	0	0	2	0	2	3	1
25	2	1	2	1	2	0	1	5	5	4
26	5	3	3	2	2	4	2	3	2	1
27	2	5	1	3	1	5	1	2	5	3
28	6	2	2	1	0	1	4	1	2	4
29	4	1	0	1	2	1	3	1	3	2
30	2	1	2	3	1	1	2	0	3	1

Exhibit 6.2 A Sample of Demand Data During the Non-Peak Season for 10 Items of the Same Product Category (Adapted from and Scaled in Units with Respect to the Real Data Set)

	Item									
Day	1	2	3	4	5	6	7	8	9	10
1	2	1	1	0	1	0	0	0	0	0
2	0	1	0	1	1	1	1	0	0	1
3	1	0	1	0	1	0	1	1	1	1
4	1	1	0	1	0	1	0	0	0	1
5	1	0	1	0	0	2	0	1	2	0
6	2	0	1	1	1	0	0	1	3	2
7	0	2	0	0	0	0	1	2	1	1
8	1	0	2	1	0	0	0	1	1	0
9	0	1	0	0	1	1	1	1	1	1
10	0	0	2	0	0	1	1	2	1	1
11	1	1	1	0	0	1	1	1	1	1
12	0	4	0	1	0	0	1	2	1	0
13	1	2	1	0	2	0	1	1	1	1
14	2	0	3	1	0	0	1	1	3	5
15	0	3	0	0	0	1	3	1	0	2
16	1	1	2	0	1	1	2	3	0	0
17	0	0	1	1	0	1	1	1	1	0

continued

Exhibit 6.2 continued

18	2	0	0	0	1	0	0	4	0	2
19	0	1	0	0	0	0	0	2	1	3
20	1	0	0	0	1	0	0	1	4	2
21	0	0	1	1	0	0	0	0	1	0
22	2	0	1	0	0	1	1	1	2	1
23	0	2	0	1	1	1	0	0	0	2
24	0	0	1	0	1	2	0	1	3	1
25	0	1	1	1	0	0	1	4	1	4
26	1	0	0	0	0	1	2	0	0	1
27	2	0	0	0	1	0	1	0	0	3
28	1	2	1	1	1	0	0	1	0	4
29	0	0	0	1	0	1	3	1	3	2
30	0	1	0	1	0	0	2	0	3	1

Exhibit 6.3 Inventory Level of the 10 Items of the Same Product Category (Adapted from and Scaled in Units with Respect to the Real Data Set)

Item	Inventory Level (Including Safety Stock Kept Every Day on Retail Sales Floor)
1	5
2	5
3	5
4	5
5	5
6	7
7	7
8	7
9	7
10	7

Exhibit 6.4 The Mean and the Standard Deviation of the Daily Demand During the Peak Season

Item	Mean	Standard Deviation
1	2.13	1.41
2	2.07	1.34
3	1.67	1.15
4	1.47	1.01
5	1.23	1.01
6	1.83	1.62
7	2.97	1.65
8	2.80	1.58
9	3.63	1.35
10	2.03	1.45

Exhibit 6.5 The Mean and the Standard Deviation of the Daily Demand During the Non-Peak Season

Item	Mean	Standard Deviation
1	0.73	0.78
2	0.80	1.03
3	0.70	0.79
4	0.43	0.50
5	0.47	0.57
6	0.53	0.63
7	0.83	0.87
8	1.13	1.07
9	1.17	1.18
10	1.43	1.30

Exhibit 6.6 The Amount of Required Safety Stock During the Peak Season with a Given Target Inventory Service Level

Item	Inventory Service Level		
	90%	95%	99%
1	1.80	2.32	3.27
2	1.71	2.20	3.11
3	1.48	1.90	2.69
4	1.29	1.66	2.34
5	1.29	1.66	2.34
6	2.08	2.67	3.77
7	2.11	2.71	3.84
8	2.03	2.61	3.69
9	1.73	2.22	3.14
10	1.86	2.38	3.37

Exhibit 6.7 The Amount of Required Safety Stock During the Non-Peak Season with a Given Target Inventory Service Level

Item	Inventory Service Level		
	90%	95%	99%
1	1.01	1.29	1.83
2	1.32	1.70	2.40
3	1.02	1.31	1.85
4	0.65	0.83	1.17
5	0.73	0.94	1.33
6	0.81	1.03	1.46
7	1.12	1.44	2.03
8	1.38	1.77	2.50
9	1.51	1.94	2.74
10	1.67	2.15	3.04

Exhibit 6.8 The Inventory Level with Different Target Service Levels

	Peak Season			Non-Peak Season		
	Inventory Service Level			Inventory Service Level		
Item	90%	95%	99%	90%	95%	99%
1	3.94	4.45	5.41	1.74	2.02	2.56
2	3.78	4.27	5.18	2.12	2.50	3.20
3	3.15	3.57	4.35	1.72	2.01	2.55
4	2.76	3.12	3.81	1.08	1.26	1.61
5	2.52	2.89	3.57	1.20	1.41	1.80
6	3.91	4.50	5.60	1.34	1.57	2.00
7	5.08	5.68	6.81	1.95	2.27	2.87
8	4.83	5.41	6.49	2.51	2.90	3.63
9	5.37	5.86	6.78	2.67	3.10	3.90
10	3.89	4.42	5.41	3.11	3.58	4.47

Exhibit 6.9 The Inventory Savings with Different Target Service Levels

	Peak Season			Non-Peak Season		
	Inventory Service Level			Inventory Service Level		
Item	90%	95%	99%	90%	95%	99%
1	1.06	0.55	−0.41	3.26	2.98	2.44
2	1.22	0.73	−0.18	2.88	2.50	1.8
3	1.85	1.43	0.65	3.28	2.99	2.45
4	2.24	1.88	1.19	3.92	3.74	3.39
5	2.48	2.11	1.43	3.80	3.59	3.20
6	3.09	2.50	1.40	5.66	5.43	5.00
7	1.92	1.32	0.19	5.05	4.73	4.13
8	2.17	1.59	0.51	4.49	4.10	3.37
9	1.63	1.14	0.22	4.33	3.90	3.10
10	3.11	2.58	1.59	3.89	3.42	2.53

Exhibit 6.10 The Inventory Cost Savings (in U.S. Dollars) with Different Target Service Level for the 10 Items

Item	Peak Season			Non-Peak Season		
	Inventory Service Level			Inventory Service Level		
	90%	95%	99%	90%	95%	99%
1–5	88.50	67.00	26.80	171.40	158.00	132.80
6–10	178.80	136.95	58.65	351.30	323.70	271.95
ALL	267.30	203.95	85.45	522.70	481.70	404.75

Case 7

Network Design at Commonwealth Pipeline Company

Matthew J. Drake, Duquesne University

Introduction

Commonwealth Pipeline Company is a major pipeline transportation provider in the Northeast, Mid-Atlantic, and Midwestern United States. The commodities the company hauls most often are refined petroleum products as well as propane, butane, refinery feedstocks, and blending components. Commonwealth operates as a common carrier in accordance with published rates and rules tariffs. Commonwealth also provides its customers with bulk storage services at many terminals throughout its service region.

Pipeline Network Repair and Replacement

As the company was founded more than 100 years ago, many of its underground pipes are in need of repair and replacement. The pipes serving its Pennsylvania terminals (listed in Exhibit 7.1) are especially worn, and Commonwealth is considering a large-scale replacement of all of its pipes in the state. The firm would like to accomplish this project at the lowest cost. Because the pipes themselves are the most expensive part of the project, Commonwealth wants to identify the best network design that uses the least amount of piping, yet still connects all the terminals in the state. Exhibit 7.2 lists the possible connections and the associated distance between pairs of terminals, and these possible connections are depicted in Exhibit 7.3.

Note that the distances in Exhibit 7.2 are the driving distances between the terminals, which are obviously different from the amount of pipe that would be needed to make the connections because the pipes are not restricted to traversing road routes. They will, however, serve as a good proxy of the required pipeline distance between the locations for the purposes of this analysis.

Your job is to determine the network design that minimizes the total mileage of pipe utilized, such that all 21 of the Commonwealth terminals are connected. Draw the arcs of the optimal pipeline network on the map provided in Exhibit 7.4.

This analysis is a simplified version of the problem that Commonwealth Pipeline Company actually faces. Complete your analysis by suggesting several practical considerations that Commonwealth would have to consider when choosing the connections to include in its network to produce the best overall network design. Can you think of a way that these considerations could be incorporated into your basic model to provide a more comprehensive network design model?

Exhibits

Exhibit 7.1 Commonwealth Pipeline Company's Pennsylvania Terminal Facilities

Node #	City #
1	Coraopolis
2	Stowe Township
3	Indianola
4	Midland
5	Reading
6	Tuckerton

Case 7 • Network Design at Commonwealth Pipeline Company

Node #	City #
7	East Freedom
8	Malvern
9	New Kingstown
10	Carlisle
11	Halifax
12	Highspire
13	Philadelphia International Airport
14	Mount Union
15	Exeter
16	Dupont
17	Lancaster
18	Allentown
19	Macungie
20	Delmont
21	Greensburg

Exhibit 7.2 Possible Connections and Distance Between Commonwealth's Terminals

Arc	Mileage	Arc	Mileage	Arc	Mileage	Arc	Mileage
(1,2)	11	(1,3)	31	(1,4)	24	(2,3)	21
(2,4)	34	(2,20)	31	(2,21)	40	(3,4)	62
(3,20)	24	(3,21)	33	(4,20)	64	(5,6)	10
(5,8)	42	(5,11)	79	(5,13)	71	(5,15)	100
(5,17)	27	(5,18)	37	(5,19)	34	(6,8)	44
(6,11)	84	(6,15)	93	(6,16)	89	(6,17)	33
(6,18)	30	(6,19)	27	(7,10)	106	(7,14)	56
(7,20)	76	(7,21)	77	(8,13)	33	(8,16)	114
(8,17)	45	(8,18)	58	(8,19)	43	(9,10)	8
(9,11)	34	(9,12)	19	(9,14)	67	(9,17)	47
(10,11)	41	(10,12)	28	(10,14)	67	(11,12)	31
(11,14)	85	(11,15)	101	(11,17)	59	(11,20)	205
(12,17)	30	(13,17)	88	(13,18)	69	(13,19)	75
(14, 20)	146	(15,16)	6	(15,17)	127	(15,18)	67
(15,19)	73	(16,18)	64	(16,19)	70	(18,19)	7
(20,21)	10						

Exhibit 7.3 Map of Commonwealth Pipeline Company's Facility Locations and Possible Connections

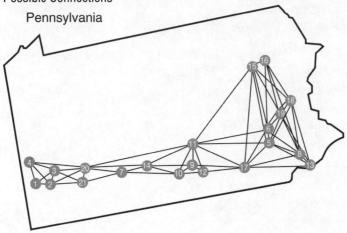

Exhibit 7.4 Optimal Pipeline Network Design

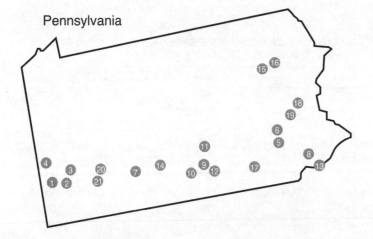

Case 8

Publish or Perish: Scheduling Challenges in the Publishing Industry

Beate Klingenberg and David Gavin, Marist College

Introduction

Imagine yourself seated at a desk with a pen and a blank piece of paper. After a time, possibly as short as a few weeks, using ordinary words and your imagination, you craft a story that touches the hearts of millions. As a result, you become a millionaire many times over. If this scenario sounds impossible, then realize that it is exactly what J.K. Rowling accomplished with the Harry Potter series. Of course, if the author becomes a multi-millionaire, just think how profitable this project was for the publisher. It seems so easy, doesn't it? If it were that easy, everybody could get rich writing books. However, not all manuscripts make it to print, and not all printed books really appeal to the public.

Now imagine that J.K. Rowling's last book is getting ready for release, and a problem occurs during printing. Eager fans are lining up at the booksellers on the release date—and the books are not there. What a disaster for the author, the publisher, and the bookstores. Fortunately, this did not happen—all customers walked away happy on the release date! So what does it take to make a book release happen smoothly? Who is working behind the scenes? The following is a case of a small publishing firm, struggling daily to keep projects on time. Unfortunately, things do not go smoothly all the time.

The Book Publishing Industry

The U.S. book publishing industry consisted of approximately 3,100 publishers[1] in 2008 and around 87,000 (mostly small ones) in 2013.[2] Industry revenue expanded 6.9% in 2012 to $15.05 billion.[3] Although e-books had an approximately 22.6% market share in 2012,[4] the printed book market remains healthy.[5] The top four publishers for trade books in 2013 are Penguin Random House, Harper-Collins, Simon & Schuster, and Time Warner.[6] The industry published approximately 329,000 books in 2012.[7]

There are a few basic realities required to understand the book publishing business. The first reality is that quality manuscripts are paramount. A quality manuscript starts with a concept or idea that appeals to a large segment of the population. Then, the idea is crafted into a story that makes the finished product hard or impossible to put down, and when people read the story, they can't wait to tell others about it. Finally, the manuscript is delivered on time to the publisher, needing minimal development or revision.

---

[1] U.S. Census Bureau, Statistical Abstract of the United States: 2012.

[2] Standard & Poor's NetAdvantage Database, 2013, www.netadvantage.standardpoor.com, accessed July 7, 2013.

[3] Ibid.

[4] Ibid.

[5] Association of American Publishers, www.publishers.org, accessed July 7, 2013.

[6] J. Lilliot. "Top Five Pubs Take Half of Sales," *Publishers Weekly*, 2005, 252(17), 5–7.

[7] Bowker, "Publishing Market Shows Steady Title Growth in 2011 Fueled Largely by Self-Publishing Sector," http://www.bowker.com/en-US/aboutus/press_room/2012/pr_06052012.shtml, accessed July 7, 2013.

The challenge of discovering quality manuscripts usually requires the talents of everyone in the organization. In many publishing companies, every employee is charged with being on the lookout for creative, unique ideas for publishing projects. Of course, the editors and executives are primarily responsible for selecting the projects and then turning those ideas into highly salable products.

The second reality is that there are two selling seasons in the book publishing industry. The first is the spring season, during which books are produced and distributed for the summer buying season. The summer buying season caters to people who have additional leisure time because of work or school vacations. The second season is the fall season, during which books are produced and distributed for the December holidays. During the fall season, books are not only purchased for personal enjoyment but also to be given as gifts. These two seasons significantly affect overall revenue.

The third reality is that books have to be created, printed, and shipped prior to the new season's start. If a book project misses the season's start, it will most likely fail to achieve the projected sales. This is a result of the flood of books hitting the market during each season.

After each season, books are backlisted.[8] After a book is backlisted, book buyers are not enthusiastic about ordering it because of the large quantity of new books coming on the market. However, a backlisted book will continue to be heavily ordered if the book experienced extraordinary sales during its initial release.

APG Publishing Company

Austin Publishing Group (APG) is a book and magazine publishing company located in Austin, Texas. The company grew out of a small advertising agency originally located in Houston, Texas. Initially, the

[8] For an explanation of this and other terms used in the publishing industry, refer to Exhibit 8.4.

company enjoyed surprising success with its early publishing projects. One of the most successful projects told the story of a local professional quarterback who had captured the hearts of fans across the country. The company produced a beautiful color pictorial book that sold tens of thousands of copies. When the book finally went out of print, it became a collector's item.

That early success caused the company to cease advertising agency operations to devote all of its resources to book and magazine publishing. After a few years, the magazine division split from the book division and formed a separate corporation. Soon both companies were thriving with very exciting futures ahead.

APG mainly produces trade books for adults; however, occasionally there are projects that cater to children. In a typical year, the number of publishing projects fluctuates between 25 and 50. The total number of projects is dictated by the number of quality manuscripts "discovered" prior to the start of the "season."

The types of books produced by APG are

- Biographies of famous people, such as politicians and business people
- Sports books about world championship teams and sports heroes
- Children's education books
- Craft, gift, and humor books
- Cookbooks
- Health and fitness books

Being located in Texas affords APG some great manuscript opportunities. The state is home to many business legends, Fortune 500 companies, high-profile sports teams, and political figures including many presidential candidates. During one particular period, APG was negotiating book proposals with four billionaires. One owned one of America's premier football teams, another owned one of the world's largest private investment companies, the third owned a very large natural resource pipeline company, and the fourth owned an international restaurant company. What contributed to the excitement was the fact that APG was able to acquire three of the four manuscripts.

The Production Department

The production department turns raw manuscripts into finished books; this is the place where everything comes together! Any problem, be it with the manuscript itself, layout, or printing, is addressed within this team. APG has just enough employees to accomplish the required tasks. The positions in the production department include the following:

- The Chief Operations Officer (COO) is responsible for managing the overall production process, which requires planning and controlling all process steps.
- The Managing Editor (ME) is primarily responsible for finding quality manuscripts or developing ideas into quality manuscripts. Additionally, this position is responsible for editing the manuscripts for content and readability.
- The Assistant Editor (AE) is primarily responsible for editing projects as assigned by the Managing Editor. The tasks can include second edits, final edits, and galley proof edits, as explained in the production process detailed in the following section.
- The Art Director (AD) is primarily responsible for designing both the exterior and the interior of the book. The interior of the book covers the type font, type size, art, callouts, and so on. The exterior of the book encompasses the front cover, back cover, and the spine; in other words, the complete dust jacket. Of course, the most important design element is the front

cover because it is the prominent selling feature of the book. The AD completes the dust jacket by combining the chosen art work with the cover marketing developed by the marketing department.

- The Assistant Art Director (AAD) is primarily responsible for the actual layout of the interior (pagination) of the book. The AAD takes the edited manuscript and creates the finished interior look according to the design and art provided by the Art Director.
- The Production Manager (PM) is primarily responsible for coordinating the various steps in the production process and for developing a production schedule. The PM ensures that projects are kept on schedule. Some of the PM's most important duties are to find the best printer for each publishing project, obtain printing quotes, and ensure that the printers deliver products on time.
- The Warehouse Manager (WM) is primarily responsible for delivering printing jobs and processing shipments to retailers. The WM ensures efficient and timely shipping and controls the inventory.

The Book Production Process at APG

During the manuscript selection process, the COO, Vice President of Marketing, ME, and PM discuss each project to determine what the final product should look like. The initial specifications become the objectives that guide the production department during the production process. Some of the specifications cover the physical size, number of pages, cover type, and interior art. These specifications are also used to determine a budget and production schedule.

Production Steps

1. **Manuscript received:** This is the starting point of the production process. The author of the manuscript sends the completed manuscript to the ME. If the author misses the agreed-upon deadline, the production process suffers.

2. **First edit:** At this stage, the editor reviews the manuscript to see whether it is worth publishing. Then a thorough reading of the manuscript is done to make sure the story is logical and complete. Also, any writing style changes are done here.

3. **Second edit:** Following the first edit, the manuscript is reviewed a second time with a focus on readability and punctuation. At the completion of this step, the manuscript is considered "clean" and ready to be flowed into the final layout of the book.

4. **Interior design and art creation:** In this step, the AD creates an interior design that matches the book's unique features. For instance, if the book is about soccer, the AD tries to pick a type font and other features that would appeal to this particular audience. The AD might choose to add soccer balls with numbers in them as the folios (page numbers). Any interior art, such as photographs or pictures, is also created. These tasks are initiated after the first edit is completed.

5. **Cover art:** The front cover art is one of the most important tasks for the AD. Typically, he creates up to a dozen different designs for each project. This task starts also after the first edit is completed, meaning that the AD is handling two important design tasks simultaneously.

6. **Interior layout (pagination):** In this step, the AAD flows the edited manuscript onto the blank pages of the book. The interior design and art are also added during this stage, meaning they have to be complete by this stage. The second edit also needs to be finished.

7. **Final proofread before printing:** After the interior of the book is completed, the book is sent back to the ME or AE for one final read-through. This is the last chance to make changes before sending the project to the printer.

8. **Cover marketing created:** The marketing department usually takes charge of developing copy for the dust jacket including back cover copy and the inside flap copy. This is a very important task because customers are motivated to purchase books through back cover and flap copy. These areas explain why the book is appealing to a particular audience.

9. **Creation of dust jacket design:** Based on the cover art and the cover marketing, the AD creates a "look" that matches the front cover art and completes the design of the dust jacket.

10. **Printer selection:** The PM starts looking for the right printer to print each particular job. At some point in the production process, a printer whose production schedule can accommodate the job has to be selected. As printers vary in their production capabilities, this decision is best made after the interior layout is completed. A printer's production capabilities are generally categorized as one-color books, four-color books, softcover books, standard size books (6 × 9 inches), oversize books, short run (fewer than 5,000 copies per run) or long run (more than 5,000 copies per run).

11. **Project sent to printer:** When the final edit and the dust jacket are completed, the book is sent to the printer by the PM.

12. **Creation of galley proofs:** After the initial work is done by the printer, a galley proof (basically a printed test version of the book) is created.

13. **Galley approval:** The ME or AE performs a final review of the book. This is the last chance to make any changes or correct possible errors.

14. **Printing:** The book is printed by the printer of choice and shipped to the warehouse.

15. **Receiving the project from the printer:** This is always an exciting event, with all stakeholders celebrating when a new project is completed and prepared for distribution.

16. **Project shipped to booksellers:** In the last step in the production process, books are delivered to the bookstores. This task is completed by the warehouse staff under the supervision of the WM.

Production Schedule

As each project is received from the author, the COO enters it onto the production schedule with data for the processing time of each step of the production process. Each week the schedule is reviewed and updated by the entire production department. During this review, production team members discuss production problems, and the team works together to resolve them.

One of the challenges the production department continually faces is how to effectively schedule the various projects. Given that each project has many unique requirements, estimating the time to complete each step is one of the first challenges. For instance, one project might have a great deal of text, which places extra burden on the editing and proofing steps. Another project might have a lot of graphics or art, which places extra burden on the AD and the AAD. Of course, inexperienced authors are one of the most common scheduling problems. They often want extra time to complete the manuscript, or worse, they submit a poorly written manuscript, which then requires extensive editing.

Typical schedules for different project types are provided in Exhibit 8.1. As APG is looking into improving its scheduling process, the COO suggests developing statistical data for each processing step.

Thus far, the team has compiled mean processing times and standard deviations for each processing step, based on the projects completed in the past three years. The scheduling times for interior design and cover art reflect that only one person, namely the AD, is considered capable to complete this task (in other words, the times are longer as the AD is working on two parallel tasks).

Usually, every step in the production process is completed by APG's employees. If the production schedule encounters difficulties, the PM can use outside resources to solve bottlenecks. These outside resources are only to be used when absolutely necessary. Over time, APG has developed a backup for most of the primary production steps, except the ones that are either impractical to outsource, such as the printer selection, or considered strategic steps for success, such as the final design of the dust jacket. Backup personnel are provided for the editing, proofing, cover art, interior design, and art steps. These outside resources are experienced professionals who work as independent contractors. Because they are independent contractors, it is hard to control when and where they work. It is also difficult to ensure that they are indeed available when needed for specific projects, as backups typically work for several publishers. However, the cost to employ these independent contractors is considerably less than what it would cost to hire them on a full-time basis. For example:

- Backup editor for first or second edit: $1,000 per project
- Backup proofreader for final proof and/or galley proof: $700 per project
- Backup cover art: $1,000 per project
- Backup interior design and art: $500 per project

Rules for Scheduling Projects

- Manuscripts are given priority if they are front-listed. Front-listed means the new projects are in the "front" of the catalog and are due out before the start of the next selling season. These are the books with which the retailers are expecting to meet their sales goals, and failure to meet these deadlines causes the retailers to lose confidence in the publisher's ability to produce and deliver the promised projects.

- Projects are ranked at the beginning of each season, with highest priority going to those projects with the best sales potential. The sales potential of each project is gauged through a survey of the national sales teams and the major book retailers. Based on each project's sales potential, the Vice President of Sales negotiates a delivery date to maximize sales revenues.

- If a project develops production problems, it is considered "derailed" (like a train that has fallen off the track) and is taken out of the production schedule. Then all remaining projects are rescheduled and moved up on the production schedule. Typical problems that derail projects include the following:
  - Manuscript delivered late from the author
  - Manuscript needs major revisions
  - Missing photos (interior artwork)
  - Missing cover quotes from famous people
  - Preface and foreword arrive late
  - Cover art requires major revision
  - Trouble completing the index
  - Late delivery from the printer

- High-priority projects can be reinserted into the production schedule if they get back on track.

- Simple projects, such as those with text only, usually go through the production process faster. These projects are used to quickly fill in the production schedule to keep everyone at maximum productivity.
- If a scheduling bottleneck occurs, an independent contractor is utilized to handle the conflicting production step within the project's budget constraints. Every project is assigned an additional amount of money that can be used to assign tasks to backup personnel if needed. The amount is estimated based on the base project budget and expected revenue. This "budget buffer" is not to be exceeded.

Reality Kicks In...

APG just secured the rights to publish one of the biggest sports books of all time. A very popular sports hero is about to break the record for the most career games played. This record has not been broken in more than 50 years! The production department is excited! The project is expected to be one of the biggest books of the year. In such a situation, timing is critically important. Finished books have to be available just prior to the athlete breaking the record. Everyone connected with the project hopes for two things to drive sales: the massive publicity from the media and consumer excitement to obtain a memento of the event.

The manuscript is to be received the next day. It is late in the evening already, and the COO, Linda Miller, is still in her office, checking and double-checking schedules and personnel availability. This project needs to be a complete success! She has estimated a budget buffer of $3,000, and in terms of estimated processing times, the book is considered a "regular book." Now she is staring at the statistics of the processing times of past projects and the actual availability of the production team members (see Exhibit 8.2). As always, everybody is booked for parallel projects, and there is no slack time. In fact, some of the team members are available for fewer days than the mean of the processing time for a given production step. Does that mean APG should hire backup resources right away? And if so, for which of the production steps should they be utilized? And how can the statistics help to guide these decisions?

Overwhelmed by the challenging considerations, the COO turns away to stare out of her window. Night has fallen, and she can see the lights of Austin's city skyline. This is always a calming sight for her, and it also reminds her of other places where she had worked until late at night, inspired by all the other busy people she imagined behind every light around her. With a smile, she recalls her late study nights when she was churning through an Executive MBA program... and abruptly she turns back to her computer. The MBA program... the Decision Science course she took...it all starts to come back to her. The statistics her team had gathered from past projects are probability distributions. With that, she should be able to build a model that can simulate whether she needs to hire backup resources.

Another big sigh, but now one of excitement! She starts Excel, rummages through her bookshelf to find the textbook used for the Decision Science course, and sets off to develop the model. "Good thing that I just bought myself a little espresso maker for my office; this will be another late night...just as during my MBA program!" she thinks as she starts building the spreadsheet model.

More Trouble on the Horizon

In the end, APG scored a big success with an on-time delivery of the famous sports book. The COO's model indeed helped greatly to make critical resource decisions. However, there is never time to rest....

The book publishing industry is subject to similar business cycles as the rest of the American economy. Wondering about a looming economic downturn, Linda Miller begins looking for ways to reduce costs without affecting the quality of products and services provided by APG. She starts with the income statements of the past two years (see Exhibit 8.3). The highest expenses after cost of goods sold are royalties, but not much can be done here—without the authors, there are no books! Perhaps APG could use cheaper materials? Intuitively, the COO knows that although this can reduce some costs, more needs to be done. Revenue is so easily lost when projects derail, even with her new scheduling model…she calls for a meeting with the production department.

"Something needs to be done!" she starts her speech. "Reviewing our income statements, I think we have questions to address. Our process seems to be inflexible, which hurts us when projects derail, resulting in decreasing profits. Where are we going wrong? Can we cut costs AND become more flexible?"

How can APG change its production process to increase efficiencies (and therefore, reduce costs) without jeopardizing product quality?

Exhibits

Exhibit 8.1 General Processing Times for Various Book Projects

Task	Processing Time, Regular Book (Days)		Processing Time, Small Book (Days)		Processing Time with Interior Artwork (Days)	
	Mean	Standard Deviation	Mean	Standard Deviation	Mean	Standard Deviation
First Edit	5.1	0.8	4.4	1.1	6.8	1.4
Second Edit	4.9	0.8	4.5	1.1	6.3	1.5
Interior Design and Art	6.6	1.1	3.6	1.1	5.6	1.4

Task	Processing Time, Regular Book (Days)		Processing Time, Small Book (Days)		Processing Time with Interior Artwork (Days)	
	Mean	Standard Deviation	Mean	Standard Deviation	Mean	Standard Deviation
Cover Art	6.5	1.1	3.7	1.1	3.4	1.1
Interior Layout (Pagination)	2.8	0.9	2	0.9	14.9	1.4
Final Proofreading	5.1	0.8	4.4	1.1	7.4	1.6
Cover Marketing	9.1	0.8	9	0.9	8.9	0.9
Dust Jacket Design	1.5	0.5	1.5	0.5	1.5	0.5
Printer Selection	1.5	0.5	1.5	0.5	1.5	0.5
Project Sent to Printer	1.5	0.5	1.5	0.5	1.5	0.5
Galley Proof Creation	7.5	1.7	7.7	1.7	10.3	1.7
Galley Proof/Edit	6.5	1.9	4.9	0.8	6.6	1.1
Printing and Shipping to Warehouse	15.1	3.1	11.5	2.3	17.6	2.2
Shipping to Stores	10.7	2.3	10.5	2.3	10.5	2.3

Exhibit 8.2 Availability of Personnel for Different Processing Steps for the Sports Book Project

Task	Team Member	Availability (Days)
First Edit	ME	4
Second Edit	ME	5
Interior Design and Art	AD	7
Cover Art	AD	7
Interior Layout (Pagination)	AAD	3
Final Proofreading	AE	5
Cover Marketing	Marketing	9
Dust Jacket Design	AD	2

continued

Exhibit 8.2 continued

Task	Team Member	Availability (Days)
Printer Selection	PM	2
Project Sent to Printer	Printer	2
Galley Proof Creation	Printer	8 (Estimated)
Galley Proof/Edit	AE	5
Printing and Shipping to Warehouse	Printer	15 (Estimated)
Shipping to Stores	WM	10

Exhibit 8.3 Sample Income Statements for APG

Austin Publishing Group Income Statement Year Ending 12-31-2012			Austin Publishing Group Income Statement Year Ending 12-31-2011		
Sales		$6,125,000	Sales		$7,589,000
Cost of Goods Sold		$1,531,250	Cost of Goods Sold		$1,897,250
Returns		$1,225,000	Returns		$1,517,800
Net Revenues		**$3,368,750**	**Net Revenues**		**$4,173,950**
Salaries			Salaries		
	Executive	$200,000		Executive	$200,000
	Sales	$200,000		Sales	$200,000
	Production	$270,000		Production	$270,000
	Warehouse	$110,000		Warehouse	$110,000
	Accounting	$150,000		Accounting	$150,000
	Administrative	$65,000		Administrative	$65,000
Rent			Rent		
	Administrative	$60,000		Administrative	$60,000
	Warehouse	$50,000		Warehouse	$50,000
Independent Contractors		$0	Independent Contractors		$12,500
Office Supplies		$25,000	Office Supplies		$25,000
Warehouse Supplies		$130,000	Warehouse Supplies		$150,000
Shipping Costs		$225,000	Shipping Costs		$250,000

Promotional Costs	$500,000	Promotional Costs	$500,000
Royalties	$1,225,000	Royalties	$1,517,800
Total Expenses	**$3,210,000**	Total Expenses	**$3,560,300**
Net Income	**$158,750**	Net Income	**$613,650**

Note: APG produced 50 new projects in 2011 and 25 new projects in 2012.

Exhibit 8.4 Industry Terms for the Publishing Industry

Back cover	Part of the dust jacket covering the back of the book that usually contains quotes from prominent people used to sell the book.
Back list	Books that have already been produced in earlier seasons.
Callout	A phrase or sentence highlighted in the margin.
Copy editor	A person who reads the manuscript for content errors.
Dust jacket	The paper cover of a hardcover book. Used to protect the book's cover and as an advertising tool.
Folios	Page numbers.
Front cover	Part of the dust jacket covering the front of the book containing the title, subtitle, and graphics used to sell the book.
Front list	The books for the new selling season.
Galley	The first copy of the manuscript after it is set in type. Used to check for errors before printing.
Gantt charts	A tool for planning and scheduling projects that graphically shows when individual steps start and stop.
Independent contractors	People who are not employees but perform work for the company on an ongoing basis.
Indexer	A person who creates the alphabetical list of important words, names, and subjects, and their location within the book.
Oversized books	Books whose size is considerably larger than the industry standard book size of six inches wide and nine inches high.
Proofreader	A person who reads the manuscript looking for spelling and grammar mistakes.
Quality manuscripts	Manuscripts that are worth publishing and that appeal to a large audience.
Selling season	The period when "front-listed" projects are heavily marketed and promoted.
Type font	A set of characters with a specific size and typeface.

3
Decision Analysis

9. Narragansett Brewing Company: Build a Brewery   101
10. Aluminum Versus Plastic: A Life-Cycle Perspective on the Use of These Materials in Laptop Computers   107
11. HealthCare's Corporate Social Responsibility Program   131
12. PaperbackSwap.com: Got Books?   143
13. Stranded in the Nyiri Desert: A Group Case Study   161

Case 9

Narragansett Brewing Company: Build a Brewery

John K. Visich, Christopher J. Roethlein, and Angela M. Wicks
Bryant University

Introduction to Narragansett Brewing Company

Narragansett Brewing Company (NBC) was originally founded in 1888 in Cranston, Rhode Island, and it experienced tremendous growth right from the start. By 1908, the company was producing 196,000 barrels (1 barrel = 2 kegs = 31.5 gallons) annually, and the rapid rate of growth continued right up until Prohibition. NBC survived Prohibition, and in 1959 production reached 1,000,000 barrels. NBC reached its peak of popularity in the 1960s with 65% of the New England market share, and Narragansett was available at 80% of the bars in Rhode Island and 100% of the retail stores. Competitive pressure began to increase significantly during the late 1960s, and in 1974 Falstaff Brewing Company of St. Louis, Missouri, purchased NBC. At the time of sale, NBC was brewing 1.27 million barrels annually. Profitability problems led to the closing of the brewery in July 1981, and the Narragansett Brewery closed its doors for the last time in 1983. Most of the brewery equipment was shipped to China, and in 1998 the once-proud brewery was demolished to make way for a retail complex.

In 2005, lifelong Rhode Islander Mark Hellendrung and a group of New England investors purchased the rights and licenses to sell and market Narragansett beer. Mark recruited the brewmaster from the former Narragansett Brewery in Cranston, Rhode Island, and in 2006 the original recipe Narragansett lager was available for sale. Narragansett's sales have grown steadily since the rebirth as both product variety and geographic spread have increased. Currently, Narragansett's beers are brewed in two locations: all bottles and cans and year-round kegs at the High Falls Brewery in Rochester, New York, and seasonal kegs in Coventry, Rhode Island. Because the majority of the beer was produced out-of-state, Mark received push-back from restaurants, bars, and liquor stores in Rhode Island about why they should sell a Rhode Island beer made in another state. Narragansett turned this around with the slogan "Buy a Gansett, Build a Brewery," which became a major marketing campaign. Building a brewery in Rhode Island would validate the success of Narragansett and help to grow sales in the entire New England region. Pressure to build a brewery in Rhode Island was also coming from new craft beer competitors, who were popping up all over Rhode Island and could lay claim to being a locally produced brew.

Mark knew Narragansett needed to brew beer in Rhode Island. Moving the entire production from the High Falls Brewery in New York to Rhode Island would be expensive and disruptive. A significant investment would have to be made in equipment to brew and package large quantities of beer. Bottled and canned products are pasteurized and have a 180-day shelf life, after which the beer will begin to taste stale and/or flat. Keg beer is not pasteurized and has a shelf life of 60 days, although at 90 days, it is still acceptable. A considerable amount of space would be required for storing ingredients and packaging materials, brewing, bottling, canning, and keg filling, as well as temperature-controlled storage for the finished product. Due to the scale that would be required to perform all brewing in Rhode Island, Mark and his team decided that the best strategy would be to operate a keg-only facility in Rhode Island. This would require a much smaller investment in a facility and equipment, while giving Narragansett tighter control over a short-shelf-life product. Control was especially critical for the four seasonal beers (Bock, Summer Ale, Fest, and Porter). With a keg-only facility, Narragansett could also compete with the local craft brewers by producing a variety of unique, short-run styles.

The Keg Facility Location Decision

Keg sales in 2011 were around 5,500 barrels, and Mark estimated that the annual production of a keg facility should be approximately 7,000 barrels with a monthly seasonal peak of 700 barrels. This would require a minimum floor space of approximately 30,000 square feet (sq. ft.), provided the space available could be utilized in a highly efficient manner. Inefficient floor plans would require a larger amount of space for operations. The minimum ceiling height needed to install the brewing equipment was 12 feet. The facility would need to include room for production, inventory storage, office space, a tasting/heritage room, and a gift shop. Mark also envisioned plant tours as a way to help promote Narragansett and build customer loyalty.

Rhode Island accounts for the largest percentage of Narragansett's sales and is supplied by two wholesalers that are both located in Cranston off Exit 14 of major highway I-95. The first distributor, McLaughlin Moran, is located at 40 Slater Road. The second, Wayne Distributing Company, is at 45 Sharpe Drive. Being close to the wholesalers would reduce Narragansett's transportation costs to deliver filled kegs to the wholesalers.

Other factors that are important to NBC included community approval, the advertising and promotion potential of the building through billboards and signage, the number of vehicles that drive by each day, and the accessibility of the building to the general public for facility tours. Because breweries emit a strong aroma that might be considered an annoyance by some residents, a good fit with the local community was critical. The cost to lease the building, as well as any costs to upgrade utilities or configure the building for a brewery, would have a major influence on the facility location decision. The estimated operating costs were also very important because a low lease cost could be offset by a high operating cost. The number and location of loading docks would also be important. Proximity to a major highway, sufficient space for a heritage/tasting/merchandise room, and possibilities for expansion also needed to be considered.

In the early winter of 2012, Mark and his team began exploring properties throughout Rhode Island to develop an understanding of the commercial real estate market in Rhode Island. They took an unstructured approach by doing online research, visiting buildings, and gathering information from local real estate agents. After a few weeks of fact-finding, Mark's team identified the three best locations (labeled A, B, and C).

Location A: 90 James P. Murphy Highway, West Warwick, Rhode Island

This 50,000-sq. ft. building on 7.8 acres has 26 foot ceilings and 16 loading docks with bumpers and levelers. The loading dock doors are 10 feet high and 8 feet wide, and can easily allow access for the brewing equipment. The lease cost is $4 per sq. ft., which equates to $200,000 a year. The insulated building is equipped with high-efficiency lighting and motion sensors, which would help keep down operating costs. Renovations and upgrades would be minimal, and there is ample room for parking cars and tractor trailers. There are 40 parking spaces in front of the building for visitors and a large lot nearby for employee parking. Interstate highway I-95 is 1.5 miles away, a four-minute drive; and a billboard on the roof would be visible to the thousands of cars that drive by each day. The building is located in a business park, so noise and odor will not be a problem. The lot has two entrances: one in the front for visitors and another one that leads to the back of the building for deliveries. The open floor plan means that a highly efficient layout that does not utilize all 50,000 sq. ft. of space can be designed. However, the vacant space can be used to expand brewing production beyond the 700-barrel maximum peak production.

Location B: 95 Grand Avenue, Pawtucket, Rhode Island

This massive 400,000-sq. ft., two-floor building was constructed in 1930, but it was renovated in 2009 and would require little in the way of utility upgrades. Located in an industrial zone, the building has 14- to 18-foot ceilings and 10 loading docks. Approximately 350,000 sq. ft. on the ground floor is warehouse space, while 50,000 sq. ft. of office space is located on the second floor. One-year leases can be signed as follows: warehouse space at 10,000 to 75,000 sq. ft. for $2.50 to $3.00 per sq. ft., and office space at 10,000 to 50,000 sq. ft. for $6.00 to $7.00 per sq. ft. This space flexibility would facilitate future expansion of the brewery. Renovations will be required to secure both the production area and the office space, and the flow between the two areas will not be smooth unless offices are constructed in the warehouse space. There is ample room for signage, sufficient parking for employees and visitors, and although the building is located only 2.5 miles from I-95, it is a 10-minute drive because of the numerous stop lights. The building is also located near residential properties, so odor might be an issue with the community. However, the mayor and town council of Pawtucket are very business friendly and have streamlined the processes required to open a business in Pawtucket.

Location C: CJ Fox Building, 2 Fox Place, Providence, Rhode Island

This historic Rhode Island building on 2.16 acres of land was constructed in the year 1900 and has great visibility, as approximately 166,000 cars drive by daily on I-95. There is easy access to both I-95 as well as I-195, which leads to a distributor in southeast Massachusetts.

The building has four floors, each at 16,125 sq. ft. and the rent is $2 per sq. ft. All four floors are vacant, so Narragansett has the option of leasing any number of floors. The ceilings are 16 feet high; plus, there are three loading docks, a single large drive-in door, and parking for 100 cars. Access is tight for a 40-foot tractor trailer, but manageable if the driver has sufficient skill. The building is fire code-compliant, is fully air conditioned, and has a security system. However, due to the age of the building, extensive renovations would be required to upgrade some utilities. Although brewing operations could be conducted on the ground floor and the offices and other rooms could be located on the second floor, expansion of brewing operations would require the use of a new floor, which would lead to costly production inefficiencies. The building is located within walking distance of the Providence Train Station, the Providence Convention Center, and the downtown area. Visitors to Providence could easily walk to the brewery for a tour and to browse merchandise. The entire area around the building is zoned commercial, so brewing odor would not be a problem with the local businesses. However, due to the proximity to the downtown area, which has a high pedestrian volume, the aroma might be a problem on humid summer days.

The Brewery Location Decision

To back up Narragansett's "Buy a Gansett, Build a Brewery" slogan, a decision needs to be made regarding the location of a keg-only brewery in Rhode Island. Mark needs to utilize one or more formal facility location methods to help him compare the three locations and decide which would best meet Narragansett's current and future needs.

Case 10

Aluminum Versus Plastic: A Life-Cycle Perspective on the Use of These Materials in Laptop Computers[1]

Ryan Luchs, Drew Lessard, and Robert P. Sroufe
Duquesne University

Introduction

In recent years, trends in the marketplace have made aluminum an important material in product design due to its light weight, durability, and recyclability. No one might know this better than Steve Kapper, Innovation Manager at Durable Aluminum, Inc. Kapper has been with Durable Aluminum for seven years and in his current role for the past four years. Kapper's current goals include expanding aluminum's market share in the consumer electronics segment. His team has identified laptop computers as a primary entry point. Laptops are a growth driver of personal computers and are mainly designed with plastic for the exterior casing. Increasing aluminum consumption in this market could provide a real revenue boost to companies like Durable Aluminum.

---

[1] Although this case is based on discussion with managers from an actual company, the name of the company and data have been disguised. Additionally, the team meeting was fictional and based partially on conversations with managers from our contact company. Thus, no conclusions should be drawn about the sustainability aspects of aluminum versus plastic; instead, the case was developed as a learning tool for future students.

However, penetrating this market further will be a difficult task. Aluminum has typically been associated with high-end laptops such as the Apple Macbook Pro series. Additionally, while the laptop still commands a price premium over a similarly equipped desktop, prices have continued their long trend downward. How can Durable Aluminum convince original equipment manufacturers (OEMs) to adopt a more costly material when many consumers continue to focus on price?

Despite the upfront cost disadvantage of aluminum versus plastic, Kapper knows that Durable Aluminum can offer OEMs real value if he can get them to focus on the long-term advantages of aluminum. Consumers continue to demand products that are more environmentally friendly, and aluminum offers potential advantages for end users as it is easier to recycle, might offer more efficient cooling, and is more durable than plastic. Furthermore, the improved durability of aluminum might provide a direct advantage to OEMs in the form of lower warranty costs. However, communicating this value proposition to OEMs and helping them profit from it is no easy task.

Kapper decided to bring a broad base of organizational stakeholders at Durable Aluminum together on a team to build a strategic plan to address this challenge. The plan will be focused on positioning aluminum as the preferred choice of raw material in laptop casing design for OEMs with regard to the environment, consumer choice, and pricing.

The Aluminum Industry

Aluminum is the second-largest metals industry in the world, with 33.9 million tons produced and $93.7 billion in sales during 2008.[2] Most aluminum is not found in its pure form but rather in several

[2] "Global Aluminum Industry Profile." Datamonitor (2009). Reference code: 0199-2004. Accessed Jan. 16, 2010.

hundred different aluminum minerals that cannot be used to produce the metal. The source of aluminum that is commercially viable is bauxite, which is comprised of 40%–60% alumina. Alumina, or aluminum-oxide, is electrolyzed to produce aluminum metal. Approximately 4 tons of bauxite is required to produce 1 ton of aluminum metal.[3]

Today, the aluminum industry is dominated by 10 multinational corporations that represent two-thirds of all the aluminum produced in the world.[4] The global aluminum industry has been strong in recent years, growing at a compound annual growth rate (CAGR) of 25.1% in sales and 7.6% in volume.[5] Its recent strength derives from global industry trends that are leveraging the metal's light weight and strength. For example, automobile manufacturers are relying on aluminum for these attributes, as consumers are expecting higher efficiency vehicles.

Looking forward, Datamonitor predicts the global aluminum industry to continue to grow at a CAGR of 5.5% in sales and 7.7% in volume during the period 2009–2013.[6]

Durable Aluminum

Durable Aluminum is a world leader in the production and management of primary aluminum, fabricated aluminum, and alumina combined through its active and growing participation in all major aspects of the industry. Durable Aluminum serves a broad range of industries including aerospace and construction.

---

[3] "How Aluminum Is Produced." Rocks, Minerals, Fossils and Earth Science Supplies. Web. Accessed Jan. 1, 2010. http://www.rocksandminerals.com/aluminum/process.htm.

[4] Ibid.

[5] "Global Aluminum Industry Profile." Datamonitor (2009). Reference code: 0199-2004. Accessed Jan. 16, 2010.

[6] Ibid.

The company is headquartered in the United States and has more than 100 facilities in 20 countries. It has 50,000 employees and is regarded as one of the world's most sustainable companies because of its innovative use of alternative energy sources for aluminum production.

Sustainability Approach

At Durable Aluminum, the company has built a strategic vision for its progress toward a sustainable business, titled *Durable Aluminum—Our Sustainable Future*. Its commitment to sustainability runs deeper within the company culture each year. The goal is to integrate sustainability concepts into key processes to make them part of the core business competencies. One of the major goals within Durable Aluminum's plan is to work to increase the rate of aluminum recycling to 75% by 2020.

Sustainability reporting and transparency are fully adopted within Durable Aluminum. Exhibit 10.1 shows a summary of recent financial results for Durable Aluminum, and Exhibit 10.2 shows the trend for an important environmental indicator—total waste generated. Durable Aluminum's first sustainability report was published in 2008. The company follows the Global Reporting Initiative G3 guidelines, the RI Mining and Metals sector supplement, and the 10 ICMM sustainable development principles.

Computer Hardware Industry

According to Datamonitor, the computer hardware industry in the United States totaled $60.6 billion in 2008, growing 5.6% over the prior year. The industry consists of three primary segments: computers, storage, and other devices. The computer segment encompasses desktop and laptop computers. Storage includes memory sticks, CD packs, hard disks, and other data storage devices. The other devices segment includes computer peripherals, PDAs, organizers, calculators, and satellite navigation systems.

The computer hardware industry in the United States has performed well in the past few years, managing to report a CAGR of 6.8% during the period 2004–2008. In doing so, the United States outperformed European and Asia-Pacific markets, which recorded CAGR of 3% and 1.1%, respectively.

Market segmentation by product type shows computer sales to be the most dominant segment of the industry with 43.1% of sales in 2008. Peripherals and devices were the next largest segment with 39.2% of sales. Storage devices capture the remaining portion of the market with 17.6% of sales.

Datamonitor predicts the computer hardware industry in the United States to continue to grow at a CAGR of 4.8% (see Exhibit 10.3) during the period 2009–2013.[7]

Competitive Landscape

The computer hardware industry is concentrated in a small number of OEMs that have well-known brand names. All are multinational companies with sales in a number of markets including the United States. The top five companies' market shares are detailed in Exhibit 10.4.

Market entry and product competition are based on strong brand recognition and reputation. Buyers have very low switching costs to competitors, but the market is resistant to new competitors because of the high brand recognition.

---

[7] "Global Computer Hardware Industry Profile." Datamonitor. Reference code: 0199-0049. Accessed Jan. 15, 2010.

The market has existed in a constant state of change, as new technology and innovation drive consumer demand. For instance, Apple shifted significant market share in the PDA industry with the introduction of its iPhone in 2006. Competitors must invest heavily into product design and development.[8] Performance attributes that are emerging relate to functionality, usability, battery life, and durability.

Integration of Aluminum in Product Design

In product design, the major OEMs are beginning to incorporate the use of aluminum cases. However, product strategies have tended to place these products into the premium lines of their brands. There are a number of excellent products that can be described as high-performing but priced at the high end of the market. For example, the case for Apple's Macbook Pro is made from a single piece of aluminum, and the final product is priced between $1,200 to $2,300, depending on screen size.

Consumer Electronics End-of-Life

In early 2010, there was a growing concern about the disposal of consumer electronics that have outlived their useful life. This concern grew from the fact that the sheer volume of discarded consumer electronic devices was enormous in the United States alone. In 2007, 2.25 million tons of e-waste was generated, with only 18% of this recycled. A significant amount of e-waste was ending up in landfills. The relatively low recycling rate for e-waste was reason for concern, but there were also reasons to be optimistic. The level of awareness of the e-waste problem was increasing, and the recycling rate was

---

[8] Ibid.

responding. Additionally, consumer electronics take-back programs were on the rise (see Exhibit 10.5). Given that the recycling rate was only 15% just two years earlier, there might be evidence that consumers are responding and are changing their behavior.[9] However, the question remained whether this increasing trend would continue.

Life-Cycle Comparison of Aluminum and Plastic Laptop Cases

In an effort to develop the value proposition for aluminum, Kapper enlisted the help of Durable Aluminum's environmental department. The following is a summary of the life-cycle data for aluminum and a common plastic that many laptop cases are made out of: acrylonitrile butadiene styrene (ABS). Kapper needed to figure out how to make sense of the data and whether he could quantify the environmental advantages of aluminum that he believed existed.

Aluminum Overview

Aluminum is the most abundant element in the earth's crust. Given the reactivity of aluminum, it is not found in nature as a free metal; instead it occurs in an oxidized state. Aluminum is strong given its light weight, and it is easy to work with because of its ductility and malleability. These properties, combined with the prevalence of aluminum, have led to its widespread use in industrial and construction activities. A downside of aluminum is the energy-intensive process needed to create it. However, aluminum can be recycled without "down-cycling," and there are significant cost and environmental advantages to using recycled aluminum as compared to primary aluminum.

---

[9] http://www.epa.gov/epawaste/conserve/materials/ecycling/faq.htm

Production of Aluminum

Most frameworks for producing aluminum include four stages: bauxite mining, alumina production, primary aluminum production, and product manufacture. Bauxite mining involves extracting the ore from the ground and moving it to a production plant. From there, the bauxite undergoes a chemical process to create alumina from the ore. The alumina undergoes an electrolytic process to separate the aluminum and oxygen, with the resulting product typically a molten aluminum. Finally, the aluminum is cast into ingots, which may undergo further processing into materials such as aluminum sheet.

Plastic Overview

Plastic has become a frequent alternative to many other materials. It has rapidly replaced wood, rubber, leather, glass, metal, and other materials in many applications from packaging to construction. The beneficial properties of plastic, including versatility and low cost, make it desirable over other materials. However, the potential impact plastic has on the health of human beings and the ecosystem must be considered as a cost of making plastic.

Production of Plastic

Plastic production begins with the raw material acquisition phase. This stage covers the activities necessary to extract raw material and energy inputs from the environment, including the transportation prior to processing.[10] The main feedstock for plastics is crude oil, which must be extracted from the ground. Crude oil is comprised mainly of carbon and hydrogen, with many other elements contained at lower percentages. The amount of crude oil needed for one pound

---

[10] Chai Hoon, Koo Chai. "A Study of the Plastic Life Cycle Assessment." Thesis. Universiti Teknologi Malaysia, 2006, p. 46. Accessed Feb. 12, 2009.

of plastic is 1.95 pounds.[11] The aggregate demand for plastics is large and growing such that 4% of global oil production is used to create plastics. The total demand for oil in 2008 was 87.2 million barrels;[12] thus about 3.5 million barrels were used as a plastic feedstock that year alone.

After the crude oil is extracted, it is separated and refined into commercially important chemicals, which are used as feedstock for further production. In the case of plastic, these chemicals are used as inputs to a polymerization process that creates the plastic resin. This resin is then further processed in operations such as injection molding to make usable product.

Acrylonitrile Butadiene Styrene (ABS)

The primary type of plastic that is used in laptop cases is ABS. The properties of ABS that make it desirable for use in computers are its hardness; stiffness and impact; and weather, temperature, and chemical resistance. It is a synthetic styrene plastic copolymer made by polymerizing styrene and acrylonitrile in various proportions, depending upon the type.

Life-Cycle Analysis for Aluminum and ABS

Exhibit 10.6 contains the life-cycle analysis (LCA) information for aluminum, and Exhibit 10.7 contains the LCA information for ABS. These LCAs are limited to greenhouse gas comparisons, as this measure is easily comparable across the alternative materials. The LCA models the primary material production, injection molding or sheet formation process, and recycling processes. The product assembly

---

[11] "Plastics & Life Cycle Analysis." American Chemistry Council/Automotive Learning Center. Accessed Apr. 8, 2009. http://www.plastics-car.com/s_plastics-car/doc.asp?CID=407&DID=1609.

[12] "Oil Consumption." The British Plastics Federation. Accessed Apr. 8, 2009. http://www.bpf.co.uk/Press/Oil_Consumption.aspx.

and product use phases are not modeled. It is anticipated that these two phases will be very similar for each material, although there might be unquantified differences in the product use phase due to differences in durability and the ability of the material to conduct heat, which might enhance cooling and lower electricity consumption of the device.

Additional pertinent information is contained in Exhibit 10.8. The cost data is available for both materials as well as for the processes needed to turn the basic material into the end product. In addition, product weight and efficiency data is available. This information is critical to get an accurate comparison between the two materials, as the weight of material needed to create a case differs across material type. Furthermore, manufacturing inefficiency will affect the end cost, as well as the greenhouse gases created.

New Business Development Initiative

As Innovation Manager, Kapper is a key member of Durable Aluminum's New Business Development Initiative. The Initiative seeks to expand the use of aluminum in new and creative applications across a broad array of industries. Generally, no idea is shot down without vetting, and no industry is off-limits. Durable Aluminum dedicates a number of full-time employees to new product development and explores many ideas simultaneously. Kapper believed that the opportunities within consumer electronics and laptop computers were vast for Durable Aluminum.

Kapper formed a team of stakeholders across the organization to collaborate on the development of a strategic plan to increase the market share of aluminum in the design of laptop computers. The Laptop Strategic Planning Team would discuss customer opportunities, product positioning, pricing and profitability, regulatory implications, marketing and communication, and logistics.

The team was assembled in July 2010, and consisted of Sarah Levens, Director of Sustainability; Bryan Baptiste, Director of Marketing; Aaron Clymer, Director of Recycling Operations; James Wirick, Director of Finance; and Mary Crigger, Director of Customer Insights. The team would meet during the following months to build a strategic plan to engage OEMs as partners and customers in the broader use of aluminum.

The kickoff meeting introduced the goals and background information, including market data, OEM briefings, and current product offerings. The team was both intrigued by and apprehensive about the challenge. Laptop computers have tight margins and cost controls for OEMs. Increasing aluminum usage in laptops was not going to be easy, but if it were, there would be no need for the Laptop Strategic Planning Team.

After the initial meeting, the team agreed to meet in a few weeks to discuss each team member's opinion on the most relevant topics to help build a base for the strategic plan. The following sections are excerpts from the follow-up meeting.

Consumers

There was no denying that the environment is on the minds of consumers more than it used to be. The real question was whether consumers were willing to pay more for an environmentally friendly product. In computers, the environment was not an attribute that OEMs had emphasized in design and marketing in the past. The issue will be new to consumers, but familiar in its application. The casing might contribute to the sustainability of a laptop if made from aluminum because of the increased recyclability. Kapper was excited with the success that Durable Aluminum has had in gaining market share in premium products, but felt that aluminum had a bigger role to play. The team discussed the issue of how consumers would view aluminum at length.

Levens supported Kapper's view that aluminum had potential:

Sustainability is definitely becoming more important to consumers these days. Products that have sustainability attributes or labels have grown rapidly over the past five years, and it seems to be reaching a tipping point. Aluminum can help create the sustainable value proposition for OEMs and thus should be presented to OEMs in this light, instead of just for its superior design attributes.

Baptiste worried how this would affect current positioning:

Consumers have come to understand that aluminum stands for premium design and now associate aluminum with high end products. They don't expect to see aluminum in value or even mainstream product lines. I'm all for increasing our sales in the consumer electronics (CE) industry, but not if it means destroying our value proposition in the high end lines. The premium line is where the margins are for OEMs, and we can command a price premium over plastic in these markets because of our design capabilities. Besides, the consumers who value sustainability are also affluent. Why not use the idea of sustainability to cement our position in premium product lines?

Wirick concurred:

Aluminum in CE is now a multi-million dollar business. We want to grow this business, but it's important for us to maintain our margins, as CE already lags more profitable businesses such as aircraft.

Crigger joined the conversation:

Our research on this area is showing some interesting trends. Sustainability now ranks in the top three of important attributes in the CE space, behind product design and cost. Consumers clearly want sustainable products, but it's not clear whether they are willing to pay for this. While Bryan is correct that wealth and concern for the environment are correlated, targeting only the wealthy misses an important demographic. Survey respondents who identified their occupations as full-time students ranked sustainability as their second-most important product attribute. The size and importance of the student market to OEMs should not be underestimated.

Kapper seemed pleased with the direction of the conversation:

I'm convinced that expansion in the laptop market needs to be a priority for us, but that it's also important that we carefully consider how our expansion affects our current relationships.

Potential Product Lines

One of the issues that has typically inhibited the greater use of aluminum in the CE market is that plastic has been the material of choice due to its low cost and the availability of skilled manufacturers. Increasing the recycle rate of CE products would help push down the overall cost of aluminum, but laptops are comprised of many components derived from multiple materials, making recycling a difficult task. To really make this happen, take-back programs will need to be in place.

Crigger started:

When customers see aluminum, it enhances their perception of design and sustainability, but they expect to pay more. Is there any way we can surprise them on the last part?

Baptiste:

I think it makes sense to limit our discussion to OEMs, where consumers see a fit between their brand image and sustainability. This brand perception will help us make the case that we can charge a price premium for products that are easier to recycle. This price premium will enable us to maintain aluminum's status in the market place.

Kapper added to the conversation:

Aluminum is going to cost more for the OEMs to create their products. There is no way around that at this point. But they might be able to make up the cost difference by reaping some benefits on the back end. Aluminum will need to be integrated into the product in such a way that it will be easier to recycle. Also, the enhanced durability of aluminum might decrease warranty costs. If we can get OEMs to set up take-back programs, this will be a win-win scenario, but how can we convince them to do this?

Crigger commented:

First and foremost, consumers want the product that maximizes performance within their price point. Design and functionality are the main drivers of performance, and aluminum clearly enhances the design aspect. Additionally, products with sustainable attributes will be perceived as higher-end designs.

Kapper listened to the conversation and realized the complexity of building the business case for aluminum in CE. He believed that greater use of aluminum in CE products benefited OEMs, consumers, and Durable Aluminum, but how could Durable Aluminum communicate that to OEMs?

Environmental Impact

Kapper then asked Levens to discuss the environmental benefits of using aluminum in laptops.

Levens summarized her thoughts:

There's no question that making primary aluminum is an energy-intensive process. However, remelting and reprocessing aluminum is far less impactful than making primary aluminum. The life-cycle assessment of post-consumer aluminum is really beneficial for Durable Aluminum. Expanding the use of recycled aluminum helps serve our long-term goals, but getting OEMs to share the same goals is the key.

I believe that we can develop two selling points for aluminum with respect to its sustainable aspects. First, for OEMs that have a strong history of environmental awareness, we can simply take the approach that aluminum is easier to recycle and thus should be incorporated into the product. Second, for OEMs without that history, making the case that regulatory reform is likely and will force their hand might get them thinking about alternative materials.

Kapper seemed pleased with Levens' analysis:

Sarah, I believe that your analysis will give our sales people some great selling points for increasing aluminum consumption.

Levens tempered Kapper's enthusiasm:

Thank you, Steve. I appreciate your kind words, but let's not get ahead of ourselves. The plastics people can counter our claims with the fact that plastic outperforms aluminum on the front end. On a pound-for-pound basis, ABS creates about 60% less greenhouse gases in its primary production. Also, it is likely that efforts to increase aluminum recycling in CE will also lead to greater recycling of plastic. Our claims will not be accepted without challenge from the plastics groups.

Kapper responded:

You are correct in pointing out the short-term benefits of plastic, but I think our sales force can do the job of getting OEMs to look at the long-term benefits of aluminum.

Pricing and Profitability

Kapper wished to move the discussion toward pricing and profitability because this was developing as a key issue for the team.

Kapper began:

Designing a line of laptop cases that use aluminum is not an issue. There are OEMs out there doing that already who have had success. But we need to bring this design more into the mainstream by showing that aluminum can drive profitability across the spectrum of product lines. We need to show the

> OEMs that it is possible to use aluminum in their product design without increasing their price points. Increased demand for their products should offset any added cost of aluminum.

Levens made her case:

> In my mind, it all hinges on the spin of this. It all depends on recycling rates and what applications the recycled materials can be used in that will help to make our case. We have to do all we can to ensure recycling rates are high and rising. There is a real benefit for us, as processing recycled aluminum is cheaper than creating virgin aluminum. This might be the key to unlocking the hidden value in this proposal for the OEM.

Realizing that reserving aluminum for a premium position was not Kapper's goal, Baptiste reinforced the need for no price increase to consumers:

> Price perception is a big deal. We are able to use aluminum to help OEMs create a premium product line, but we cannot propose that an OEM price its entire line of products above the market based on an aluminum case. It will not fly. There are large groups of consumers who will not pay for the enhanced styling aluminum can offer.

Thus, the team was on board with the notion that they needed to find a way to make aluminum attractive to OEMs without the need to pass along a large price increase to consumers. However, a split remained among the group as to how to price the product to the OEMs. Should they price their cases at a premium and make the case to OEMs that aluminum's superior design capabilities would drive sales? Or should Durable Aluminum count on increased recycling rates in the future, which would drive down its own production costs?

Communication Strategy

The group also needed to decide to whom they should direct their value proposition. Although Durable Aluminum was a B2B company, consumers could clearly drive material preferences and thus their own sales.

Levens explained her thoughts:

The consumer needs to hear the core of this message first and foremost. We've seen other companies promote core competencies to consumers when they are only indirectly selling to them (such as GE Ecomagination or Intel Inside). Durable Aluminum might have an opportunity to do the same.

Clymer made the case for a more focused approach:

Participation in recycling is critical. There has to be clear and concise messaging in stores to ensure that consumers get their CE products back into the system. We should focus our message on recycling at the point of sale, not in expensive television campaigns.

Wirick added the financial perspective:

If we are talking about moving aluminum from premium to mainstream lines, I'm not sure the price points will support a large communications campaign. My thoughts were that we would need to decrease the investment in marketing for this product line, not increase it.

Kapper felt that a case for increasing marketing spending could be made to management:

This has the potential to be a high profile project for Durable Aluminum. We might be able to make the case that an increased marketing budget is warranted, as getting consumers to see aluminum as sustainable might have spillover effects to other product categories.

Logistics

Finally, the group discussed the logistics of an enhanced recycling program and how this would be the key to creating the value proposition for aluminum.

Levens began:

Participation is key! Consumers have to be involved, and we need high take-back rates for the numbers to work here. We want to make an argument that aluminum will cost less over the long term in both nominal prices and environmental impact.

Clymer added a dose of reality:

Take-back programs have been tried in the past with spotty success. There are too many layers to the reverse supply chain for there to be high participation. You need partnerships with OEMs and retailers, as well as high participation from consumers for success.

Third-party providers can be part of the equation for disassembly and sorting of materials. Everyone at the table can win and save money from recycling and reprocessing at least one material input.

Kapper wondered what Durable Aluminum's role should be in take-back programs:

We are not well positioned to start take-back programs on our own because we lack a direct connection with consumers. How can we increase the success of these programs given their strategic importance for our business?

The group had used up its allotted meeting time and thus the meeting drew to a close. Although the discussion was good and gave Kapper and the others many things to ponder, a strategy needed to be developed and time was running short.

Case Challenge

Many at Durable Aluminum felt that this was not the time to aggressively pursue the CE market. In mid-2010, the U.S. economy was fragile and it seemed that consumers were not positioned to lead the recovery. The economy was just emerging from a particularly long and deep recession, and most data pointed to consumers improving their personal balance sheets rather than spending on new goods. In this environment, did it make sense to try to sell into markets where aluminum had no immediate cost advantage?

However, at the same time, many consumers were thinking long and hard about their impact on the ecosystem. A 2008 study of consumer concerns showed that there was not only a worry about the environment, but many would actually pay a price premium for a product that was more environmentally friendly.[13] Could Durable Aluminum help OEMs find a way to educate consumers on the benefits of aluminum? Moreover, do consumers even need to be involved, or could a value proposition be articulated that could benefit OEMs directly without consumer buy-in? Finally, Durable Aluminum needed to decide on how to price the aluminum to OEMs. Should it continue to price the aluminum at a premium to OEMs, making the case that aluminum's design and sustainability aspects will drive volume, which will increase OEM profits? Or, should Durable Aluminum price the cases competitively and bank on lower material costs through increased recycling rates?

Time was running short for Kapper. He had two life-cycle analyses from two different industry groups, both showing the benefits of their own material. How could he make sense of these numbers? Did they support his belief that aluminum offered long-term benefits over plastic? Or was he being overly optimistic about consumers changing their behavior with respect to recycling? One thing was for sure: Kapper knew that if Durable Aluminum was going to make progress in this market, it would need to happen soon. Three of the five largest PC OEMs were doing complete laptop line redesigns in the next 18 months. If Durable Aluminum did not make its case now, it might need to wait years for its chance.

[13] S.M.J. Bonnini, G. Hintz, and T.J. Mendonca. "Addressing Consumer Concerns about Climate Change." *The McKinsey Quarterly*, 2008 (March), 1–9.

Exhibits

Exhibit 10.1 Durable Aluminum Income Statement in Millions of Dollars

	2009	2008	2007
Sales	18,203	19,739	22,049
Cost of Goods Sold	15,075	15,923	16,573
All Other Expenses	3,335	2,934	1,675
Total Costs and Expenses	18,410	18,857	18,248
Net Income	−207	882	4,161

Exhibit 10.2 Durable Aluminum Waste Minimization*

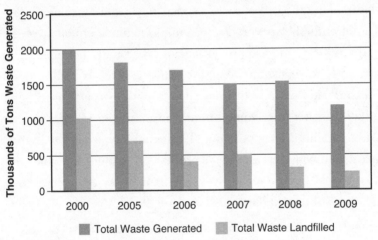

*Does not include bauxite residue.

Exhibit 10.3 Computer Hardware Industry Sales 2004–2013

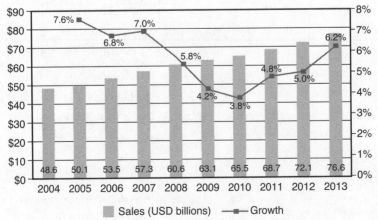

Data source: "United States–Computer Hardware" *Datamonitor*, August 2009.

Exhibit 10.4 2008 and 2009 Worldwide Top Five PC OEM Market Share

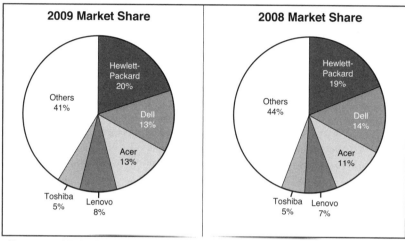

Data source: Wilkins, Matthew. "Acer Ascends, Dell Dives in 2009 PC Market." *Market Intelligence iSuppli*. March 9, 2010. Accessed July 15, 2010.

Exhibit 10.5 Current Consumer Electronic Take-Back Programs

Company	Program Name	Partnered with Outside Source?	Cost to Consumer	Process to Recycle
Apple	Apple free recycling program for computers and monitors	No	Free recycling of iPods and cell phones Free recycling of one computer and one monitor from any manufacturer	Apple sends instructions for packaging old equipment and shipping to the recycler
Best Buy	Free electronics recycling pilot program Consumer electronics recycling program	No	Pilot program: Free recycling for two pieces of equipment per day to any participating location Electronic program: Free recycling of old TVs or appliances with purchase of new Best Buy product	Drop-off kiosks Recycle phones by mail Best Buy Trade-In Center determines value Gives consumers gift card worth the value of product

continued

Exhibit 10.5 continued

Company	Program Name	Partnered with Outside Source?	Cost to Consumer	Process to Recycle
Dell	N/A	Yes: Staples	Free, no purchase required	Dell equipment drop-off at any Staples location
Gateway	Gateway trade-in and recycle program	Yes: Rechargeable Battery Recycling Corporation (RBRC)	Free recycling with purchase of new Gateway product	Value the product with Trade-In Estimator Pack and ship to Gateway Check will be issued
HP	HP asset recovery services HP's product recycling program	Yes: RBRC	Asset recovery: Free recycling of any product (asset recovery program) Product recycling: Cost of automated, online computer hardware recycling $13–$34 per item	Get quote online Shipped to HP
LG	LG Electronics Recycling Program	Yes: Waste Management	Free for LG products Charge for any non-LG product	Drop off to one of 160 Waste Management eCycling Centers

Data source: http://www.productstewardship.net/productsElectronicsBizProgramsTakeback.html.

Exhibit 10.6 Aluminum LCA Data

Aluminum Notebook Case Production

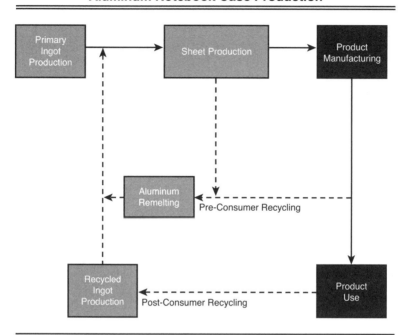

Aluminum Greenhouse Gas Data	
Process	Kg CO2-Equivalent/1000 Kg Aluminum
Primary Ingot Production	9,677
Recycled Ingot Production	506
Sheet Production	644
Aluminum Remelting	317

Exhibit 10.7 ABS LCA Data

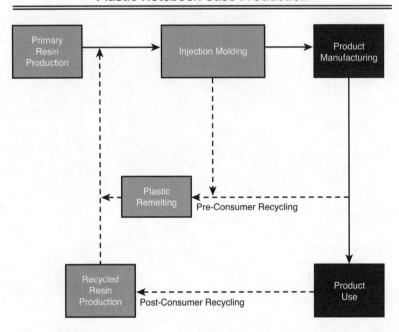

Plastic Greenhouse Gas Data	
Process	Kg CO2-Equivalent/1000 Kg Plastic
Primary Resin Production	3,760
Recycled Resin Production	3,008
Injection Molding	1,189
Plastic Remelting	1,189

Exhibit 10.8 Manufacturing Process Data

Process Data

Material	Primary Production ($/lb.)	Recycled Material ($/lb.)	Sheet Production ($/lb.)	Forming into Laptop Case ($/Case)	Assembly Cost* ($/Case)	Maximum Recycled Content in Case %Recycled Content	Material Weight (lbs./laptop)	Process Efficiency** lbs. input/lb. output
Aluminum	$ 0.99	$ 0.40	0.70	$ 5.31	Unknown	100%	2.10	1.42
ABS	$ 1.24	$ 1.00	Not Applicable	$ 6.59	Unknown	20%	1.70	1.33

*Same for both materials
**Assume material lost to inefficiency is recovered in the recycling process at the value of recycled material

Case 11

HealthCare's Corporate Social Responsibility Program

Robert P. Sroufe and Marie Fechik-Kirk,
Duquesne University

Sustainability Coordinator's First Project[1]

From her tenth floor office, Stephanie Meyers watched as the first snow began to fall over the city and thought about how much had changed over the past few months. Stephanie had gone from leading an internal auditing team at HealthCare, one of Pennsylvania's largest health insurance companies, to working independently within the facilities department as the company's first sustainability coordinator. It was Stephanie's responsibility to successfully roll out sustainability initiatives to the staff and to select the first sustainable, eco-friendly, and green project at the Pittsburgh office. The success of these sustainability initiatives was crucial, as Stephanie was now evaluated based on the success of the initiatives and on the projects she instituted. Stephanie also wanted this first project to build awareness and buy-in.

Stephanie knew there was some internal interest in the company, but she was nervous about selecting the first sustainability project.

---

[1] Preparation of this case is for the basis of class discussion rather than to illustrate either effective or ineffective handling of an administrative situation.

After eight years in accounting, Stephanie knew how to evaluate the financial, operational, and reputational risk of a project or procedure; but she wanted her first project to offer the staff at HealthCare tangible benefits. Stephanie knew she could win over more people if the community and her colleagues at HealthCare believed they benefited from sustainability initiatives. Stephanie's goal was to select a project that would conserve energy and resources while benefiting HealthCare both economically and socially. Any building renovation project needed to reduce energy usage, decrease operating costs, and benefit the environment and the community.

For the past month, Stephanie had been researching ideas to upgrade HealthCare's 30-story headquarters. Built in 1988, the building consists of 2 floors of retail space, 28 floors of office space, and a 3-level underground garage. HealthCare is the primary tenant of the building and is exploring the possibility of seeking a LEED for Existing Buildings: Operations and Maintenance certification. After careful consideration of the options, Stephanie had narrowed her decision to three possibilities: HVAC replacement, exterior tree replacement, or interior tile replacement. From her experience in accounting and internal auditing, Stephanie knew that building a strong business case for any new project was crucial. However, after going through the information, Stephanie could create a business case for all three projects to varying degrees. Yet, facilities management would make the final decision, as they were responsible for completion of the project. Looking out into the falling snow, Stephanie decided to call Gary Render, the Contracting Solutions Leader of Santoro, a leading commercial HVAC provider in Pittsburgh, to get his input. Stephanie had met Gary at a Green Building Alliance sustainability workshop last year. Gary was well connected and knew the facilities director well, and might offer Stephanie some insight.

The phone rang several times, and as Stephanie was about to hang up, she heard "Hello, Gary Render," on the line. Grateful for the opportunity to get his input, Stephanie told Gary about her upcoming decision. However, much to Stephanie's dismay, Gary did not offer her a solution. "Stephanie, HealthCare selected you to take on this task because you have the internal knowledge and the understanding of sustainability to make the best choice for HealthCare." Gary continued, "Instituting change is like driving a truck down a winding hill. You have to go slowly and carefully to avoid tipping over. You don't want to pick anything too extreme, or you'll lose people. Similarly, you also have to tie down everything carefully, so that you keep control of your load. When you present your ideas to the committee, prioritize and carefully defend your recommendations. Show that the initiative you select will save HealthCare money and will demonstrate good corporate citizenship and best practice within the industry." Gary also suggested working with a local university to have a team of business students help to confirm the analysis and opportunities. This kind of experiential learning is beneficial to both parties. Plus, community involvement aligns with HealthCare's corporate vision and values.

As Stephanie hung up the phone, she looked over the numbers again. Perhaps she needed to look not only at the numbers but also at the alignment between the project and HealthCare's mission (see Exhibit 11.1). No matter which project she selected, Stephanie needed to explain why HealthCare should complete the project and how the project aligns with HealthCare's commitment to promote and apply sustainable, eco-friendly, and green business practices. Stephanie also needed to consider the interests of multiple stakeholders, so that her project recommendations would be accepted not only by the finance department but also by the facilities department and others working in the building.

HealthCare's Current Situation

HealthCare is one of the largest health insurance companies in Pennsylvania with 4.6 million people served by HealthCare's programs. This 70-year-old company is a major employer in Pennsylvania and West Virginia, employing 12,000 people in Pennsylvania and 19,000 overall. See Exhibits 11.2 and 11.3 for the company's financial performance and balance sheet information. HealthCare's corporate office is located in Pittsburgh, Pennsylvania, and HealthCare's offices are situated in two buildings in downtown Pittsburgh. Stephanie's office is in the older 30-story Twelfth Avenue Place Building. The 86-year-old building houses not only office space but also a customer walk-in center, an employee fitness center, a 300-seat auditorium, and a center for grieving that supports children, parents, and families throughout southwestern Pennsylvania. The building served as a department store until 1995, when HealthCare became the primary tenant. HealthCare also has call centers in Pittsburgh and three other Pennsylvania locations. In 2007, these call centers received more than 6.7 million customers and provider inquiries. HealthCare's computing infrastructure is provided by a Silver LEED-certified data center, also located in Pennsylvania.

HealthCare's vision involves being a leader within the industry and meeting the health care needs of members, while maintaining a strong bottom line. In 2007, HealthCare captured $12.4 billion in revenues and dedicated $137 million to support its mission of providing "access to affordable, quality health care, enabling individuals to live longer, healthier lives," and $6.3 million to 1,800 non-profits serving Pennsylvania by providing services that help people to live longer, healthier lives.

Since July 2008, Stephanie Meyers has been coordinating HealthCare's Corporate Social Responsibility program. This program supports both HealthCare's vision and mission. Conserving resources and providing a healthy work environment for employees reduce operating costs and increase productivity. Corporate social responsibility also benefits the community by reducing emissions and reducing the impact on the local environment. According to Gary Render, corporate programs aligned with sustainability also help to mitigate the health risk associated with poor air quality, which aggravates asthma and other respiratory problems. In Pennsylvania, 60% of electricity is produced by coal-burning power plants.[2] These power plants produce particle pollution, which is one of the most dangerous outdoor air pollutants. Breathing particle pollution year-round can shorten life by one to three years and can aggravate respiratory disorders and trigger heart attacks or strokes.[3] In the 2008 State of the Air report, the American Lung Association noted that Pittsburgh had the second-highest level of year-round particle pollution and the highest level of short-term particle pollution in the country.[4] In the long term, being better stewards of the local environment could help HealthCare to reduce health care costs. Reducing emissions is particularly important for children. Studies show that children exposed to air pollution often face reduced lung function, increased incidence of asthma, and increased visits to the doctor's office and emergency room. Additionally, infants and unborn children are also at risk. Exposure to power plant pollutants can lead to low birth rate, increased incidence of premature birth, and stunted lung development.[5]

HVAC Replacement

A building's heating, ventilation, and air conditioning system (HVAC) controls the indoor climate and accounts for 40%–65% of the energy used in commercial or institutional buildings. Large office buildings require some cooling throughout the year to reject heat emitted from lighting systems, equipment, and employees. HVAC systems also allow large buildings to meet mandatory ventilation

---

[2] http://www.cleanair.org/Energy/energyImpacts.html

[3] http://www.stateoftheair.org/2008/air-basics/

[4] http://www.stateoftheair.org/2008/most-polluted/

[5] Bruce L. Hill, and Martha Keating. "A Clear the Air/Physicians for Social Responsibility Report: Children at Risk: How Air Pollution from Power Plants Threatens the Health of America's Children." (Boston: Clean Air Task Force, 2002). 1–3.

requirements. Poorly designed or outdated systems can lead to occupant distraction, loss of business, or even lawsuits.[6] Upgrading or replacing an older, less efficient HVAC system can reduce energy costs[7] and improve indoor air quality. Improving indoor air quality through upgrading or retrofitting an HVAC system can lead to better work performance, reduced medical care costs, lower employee turnover, and lower cost of building maintenance.[8] In a national survey of large office buildings, it was found that employee salaries are on average 72 times higher than annual energy costs, so a 1% improvement in productivity due to a more comfortable indoor environment could offset a building's annual energy costs.[9]

HVAC system manufacturers can provide commercial customers with custom-designed systems that maximize energy efficiency while meeting environmental and cost concerns.[10] To select the best replacement HVAC system, HealthCare will need to identify the base consumption of the current system and compare it to the consumption of new models under consideration. A decision tool incorporating assumptions and cost savings would allow for a quick comparison of the old and new systems. For example, HealthCare spends about 8.7 cents for every kilowatt-hour of energy consumed, and the current HVAC system accounts for about 65% of the building's energy costs. See Exhibits 11.4 and 11.5 for HVAC information. At the end of 2011, HealthCare's fixed energy contract with a local energy company will expire. However, costs are expected to rise, as there is talk of deregulating the state's utilities. Stephanie knows management is

---

[6] http://www.aceee.org/ogeece/ch3_index.htm

[7] http://www.fypower.org/com/tools/products_results.html?id=100124

[8] Olli Seppänen, and William J. Fisk, "A Model to Estimate the Cost-Effectiveness of Improving Office Work through Indoor Environmental Control" (ASHRAE Transactions, 2005). 663–672.

[9] Joseph J. Romm, and William D. Browning, *Greening the Building and the Bottom Line: Increasing Productivity through Energy-Efficient Design* (Snowmass, Colorado: Rocky Mountain Institute, 1998).

[10] http://www.trane.com/Corporate/About/solutions.asp

concerned about changing prices and fluctuating interest rates. It is also important to note that the HVAC system is housed on top of the 30-story building, and there would be costs in excess of $50,000 for the engineering and delivery of HVAC components by helicopter.

Exterior Tree Replacements

Stephanie's preliminary research shows healthy trees can lower summer temperatures around them, reduce storm water runoff, and improve air quality. Additionally, healthy trees can increase real estate values and increase business income.[11] In regions like Pittsburgh, which are prone to strong winds and long periods of rain, combined sewer overflow systems collect rain water and waste water. During periods of intense rain, waste water treatment centers are overwhelmed by the increase in volume and might reach their capacity. To reduce pressure on the system, untreated waste water is released directly into rivers or streams. This can lead to contamination and erosion of streams and waterways. Properly selected and planted trees can reduce storm runoff by taking in excess water. Reducing runoff alleviates some of the pressure on waste water treatment centers. Additionally, trees can reduce wind speed, which decreases the amount of cold outside air penetrating the building.[12] As HealthCare is interested in increasing sustainability, it is important to consider transportation costs and to prioritize native trees during the selection process. However, the maintenance of the trees is also critical. Trees require more care during their first few years of development. For instance, new trees should be watered, mulched, and pruned.[13] Stephanie needs to compare trees against certain criteria to select the best option. For more information about tree replacement options, see Exhibit 11.6.

[11] http://www.fs.fed.us/psw/programs/cufr/products/cufr_672_PressReleaseAlbuquerque10-06.pdf

[12] http://www.fs.fed.us/psw/programs/cufr/products/cufr_189_gtr186b.pdf

[13] http://www.clemson.edu/extfor/urban_tree_care/forlf17.htm

Interior Tile Replacement

Commercial flooring is becoming greener due to customer demand. Manufacturers are increasing the recycled content of products, reducing water and electricity usage, and increasingly manufacturers are relying on alternative energy to fuel operations. Consumers benefit from these processes because greener products not only meet performance criteria but also reduce maintenance and improve indoor air quality.[14] Alternatives to ceramic tile include bamboo, cork, or bio-based flooring. Bamboo is a renewable resource that is hypoallergenic, durable, and resists moisture well. Cork flooring is manufactured from the bark of living trees and is non-slip, comfortable, and sound-reducing.[15] Bio-based products can be made from rapidly renewable, nationally grown plant materials. The result is a tile category that offers greener flooring options along with performance features that are superior to ordinary vinyl composition tiles. For tile option information, see Exhibit 11.7.

Due to normal wear and tear, the 50,000-square foot main floor of the headquarters office building needs to be re-tiled. HealthCare would like to select the most environmentally friendly flooring product possible to replace the existing ceramic tile in this high traffic area. However, the financial impact of the new flooring is still a major consideration, as there is information in the popular press about rising costs of health care and potential economic problems at a national level. Stephanie needs a decision tool to help determine the best replacement option utilizing both qualitative and quantitative inputs.

Given this initial set of opportunities, each with significantly different financial and timing implications and criteria for decision

---

[14] Michael Chmielecki, "Commercial Flooring Asks, Are We 'Green' Enough Yet?" *National Floor Trends*, 2008, 10(11), 10–12.

[15] Marianne Wilson, "Eco-Friendly Flooring." *Chain Store Age*, 2006, 82(9), 78.

making, Stephanie needs to choose a project and present her rationale and implementation framework to management and the facilities department. Stephanie has engaged a local university's business school and a team of students to help guide the decision-making process while drawing from their business analytics toolkit.

Exhibits

Exhibit 11.1 Information from HealthCare's Website

We demonstrate our vision and mission through the innovative products, programs and services we provide, and our community involvement.

Our values:

- United in our mission
- Dedicated to excellence
- Committed to integrity
- Focused on our customers, employees and communities.

HealthCare's diversity and inclusion initiative

Through our diversity initiative, we strive to capitalize on the strengths of our many differences and the advantages of an inclusive workplace.

Sustainable, eco-friendly and green business practices

At HealthCare, we know that creating a healthy environment helps to create healthier people. To this end, we are committed to implementing sustainable, eco-friendly and green (SEG) business practices because it is entirely consistent with our mission to help individuals live longer, healthier lives.

We are committed to integrity

At HealthCare, we value integrity above all else. HealthCare's Integrity Process and values set the ethical tone for conducting business and create a corporate culture that enhances the reputation of the company. Our Code of Business Conduct points us down the right path. Our employees use the basic Principles of Integrity as a primary road map to be successful at HealthCare and in life.

Exhibit 11.2 HealthCare's Income Statement

Years Ended December 31	2007	2006
Revenue*		
Premium Revenue	$10,252,073	$9,433,837
Management Services Revenue	$621,555	$685,182
Vision Revenue	$963,119	$629,709
Net Investment Income	$235,400	$226,363
Net Realized Gain on Investments	$159,386	$25,910
Gain on Sale of Business Interests	$4,268	$42,059
Other Revenue	$116,759	$41,193
Total Revenue	$12,352,560	$11,083,803
Expenses		
Claims Incurred	$9,044,986	$7,896,422
Operating Expenses	$2,679,362	$2,432,118
Interest Expenses	$68,440	$51,888
Total Expenses	$11,792,788	$10,470,428
Income Before Income Taxes	$559,772	$613,375
Income Tax Provision	$184,410	$215,085
Net Income	$375,362	$398,290

*All numbers in thousands

Exhibit 11.3 HealthCare's Balance Sheet

	December 31, 2007
Assets*	
Cash and Cash Equivalents	$1,018,741
Accounts Receivable	$6,242,669
Trade	$977,217
Government Programs	$266,976
Other	$65,817
Property and Equipment, Net	$354,397
Total Assets	$8,925,817
Liabilities and Reserves	
Debt	$799,542
Total liabilities	$4,948,610
Total reserves	$3,977,207
Total liabilities and reserves	$8,925,817

*All numbers in thousands
Source: HealthCare.

Exhibit 11.4 HealthCare's Energy-Saving Project Assumptions

2007 kWh Consumption	25,137,000
% of Electricity Allocated to HVAC Operations	65%
Electricity Costs/kWh (out to 2012)	8.7 cents
Estimated Inflation Rate	4%
% Increase in Material and Supplies	2%
% Increase in Chemical Treatment	2%
% Increase in Small Tools and Equipment	2%
Preferred Rate: "Net Present Value"	9%

Source: HealthCare. The period of time for the analysis is over 20 years, starting with 2007 data.

Exhibit 11.5 HealthCare HVAC Expenses

Expense Description	2007 Actual
HVAC Materials and Supplies	$43,270
HVAC Chemical Treatment	$14,910
HVAC Small Tools and Equipment	$2,517
Yearly Motor Replacement	$80,000

Source: HealthCare. Any new HVAC system will not require motor replacement for eight years after the installation and will have at least a 15% improvement in energy conservation.

Exhibit 11.6 HealthCare Tree Replacement

	American Dogwood	Service Berry	Red Maple	Sugar Maple
Capital Costs (12')	$225	$245	$270	$180
Shade	Average	Minimal	Average	Average
Characteristics	Strong, highly resistant to winds, heavy rains, average growth rate	Naked oak with snowy white flowers, edible fruit; one of the first to flower in PA	Fastest growing maple tree; disease resistant, strong tree that resists ice damage and high winds; grows best in moist soils	Very large and prefers moist, rich soils; can mature in winter (maintain leaves); very resistant to high winds

Source: Information gleaned from student research and input from HealthCare.

Note: All tree replacement options are native to Pennsylvania. HealthCare is looking to replace 21 trees around the perimeter of the building coinciding with sidewalk replacement, with a total cost range of $4,500–$5,000 per tree.

Exhibit 11.7 HealthCare 50,000-Square Foot Tile Replacement

	Bamboo	Terrazzo-Concrete	Vinyl Composition Tile (VCT)	Ceramic Tiling	Bio-Based Tile
Total Cost Range	$3.75–$6.50 commercial grade sq. ft	$2.00–$5.00 sq. ft	$1.50–$2.75 sq. ft	$1.50–$2.50 sq. ft	$2.50–$3.00 sq. ft
Maintenance Cost	Medium	Low	Medium	Low	Medium
Durability	Average commercial grade	40–80 years	High (data not available)	40–80 years	Very good*
Green Attributes	Almost all have formaldehyde binders and are mostly shipped from China	Has low VOC emissions	Be wary of sealants and adhesives, as they emit VOCs	Requires no VOC-emitting products	Made with rapidly renewable ingredients**

Source: Information from MBA team's online research.

*According to Armstrong Flooring. 2× the indent resistance of VCT and 5× the impact resistance.

**Helps reduce the reliance on petroleum and fossil fuels with 10% pre-consumer recycled content.

Cork not available from local vendor at this time.

Case 12

PaperbackSwap.com: Got Books?

Brandy S. Cannon and Louis A. Le Blanc, Berry College[1]

Introduction

Many Americans have an ever-growing number of unwanted books and limited options to get rid of them. They can throw the books away if they can bring themselves to do so. Their titles can be taken to a used book store, receiving little cash in return, or given away to a friend or charity. PaperbackSwap.com has created another option by letting members trade their unwanted books with other members across the country for only the cost of postage. With a large inventory of books, hard-to-find titles are easier to locate than with a traditional bookstore.

PaperbackSwap.com began as a very small company based in Suwanee, Georgia (an Atlanta suburb). The enterprise experienced tremendous growth and amassed many loyal customers. However, with any substantial increase in the price of media mail and the rising popularity of electronic books (e-books), PaperbackSwap.com faces real threats to its model. Co-founder Robert Swarthout expressed his concern to the company over the rising prices of media mail and the uncertain future of the company. He wondered if the company would be able to survive the coming threats.

[1] Brandy S. Cannon and Louis A. Le Blanc of the Campbell School of Business at Berry College prepared this case as the basis for class discussion rather than to illustrate either effective or ineffective handling of an administrative situation.

Company Background

Robert (Bobby) Swarthout came to Berry College in northwest Georgia from Florida in 2000 to run track and pursue a degree in computer science with a minor in business. He attended Berry from August 2000 through May 2004. In 2002, during the summer break before his junior year, Swarthout had his wisdom teeth removed. He decided to create a website while he was at home recuperating and had some free time. He did not like the way that the Student Government Association (SGA) ran its book swap at the College. The SGA book swap had no Internet presence. Also, students did not know what books SGA had in stock until the day it opened, which was usually the first day of class.

Two weeks before classes began for the fall 2002 semester, Swarthout placed 250 flyers on the Berry campus advertising a website, www.berrybookexchange.com, that he had created for students to swap textbooks. The website allowed students to post their textbooks free of cost, and students purchasing the books did not have to pay any fees. Within a few days, all his flyers were down. The College's newspaper wrote an article about the incident:

> Dean of Students Tom Carver ordered signs for www.berrybookexchange.com removed from dorm halls because they were viewed as solicitation without permission. Swarthout went directly to Carver when he heard about the removal of the fliers. Carver said that when he had all the facts, he was receptive to the idea of a website,[2] although he worried that it would unfairly sway students from the book swap, taking money from SGA.

SGA made an estimated $10,000 a semester from its book swap operation.[3] Swarthout's website threatened SGA's business. Swarthout

---

[2] Lindsey Quirk, "Fliers for Book Exchange Website Taken from Dorms," *Berry College, Campus Carrier*, (September 5, 2002) 94(2), 3.

[3] Ibid.

had to talk to Dean Carver and then SGA. SGA asked him to give it control of his website, which Bobby refused to do. He later had a meeting with the Berry College administration, including then-President Scott Colley and Dean Carver. He was threatened with being expelled from school because he had allegedly used Berry's resources to create the website. However, Swarthout had created the website at home, and it was hosted off campus.

SGA then offered to buy the website. Swarthout had taken an estimated 25% of used textbook sales away from SGA.[4] Swarthout's website had no revenue model, and the students were the ones saving the money. SGA bought the website for $2,000 and assigned 25% of revenues to Swarthout for four subsequent semesters.[5] Also, Swarthout entered into an agreement that called for him to be paid $12.50 per hour to build a text swap website for SGA. It was to be ready for the spring semester 2003,[6] at which time the original site was closed.

Swarthout convinced the College to sign a contract, and he intentionally left out a non-compete clause. The following summer (2003), he built another website better than his original. In the fall of 2003, his senior year, Bobby was not allowed to advertise his new site on the Berry campus. He then created websites to swap textbooks for other schools including Shorter University (in Rome, Georgia), the University of Tennessee (Knoxville campus), the University of Mississippi, Florida State University, Clemson University, as well as the University of Kentucky. At Florida State, 3,500 people used the site.[7] No cash was generated at any of the schools as the model was essentially a prototype.

[4] Ibid.
[5] Ibid.
[6] Ibid.
[7] Ibid.

Seed Capital and Advice

Swarthout attended an alumni entrepreneurial presentation during Berry College's Mountain Day weekend (that is, homecoming) sponsored by its Campbell School of Business in October 2003. Richard Pickering, a Berry alumnus with an MBA from the Harvard Business School, gave a presentation on starting new business enterprises. While speaking to the group of students, Pickering offered to meet with any students interested in new business opportunities. Pickering offered that, if the students wished to discuss their ideas, he would potentially invest in their idea and help them build a company. Pickering assisted Swarthout to overcome the challenges and obstacles that young business leaders face.[8]

Bobby also participated in the Students In Free Enterprise (SIFE) competition with his college book swap business model. He did well in the regional competition and advanced to the national competition in 2003.

Pickering and Swarthout began talking daily about the collegiate book swap websites during the fall of 2003, sometimes several times a day. They met again in the spring of 2004. Swarthout then visited Pickering in Atlanta to brainstorm further.

Initial Operations

By this time, Swarthout had expanded the online book swapping model to 12 schools. They were now charging a $1.25 to post a book at three large universities. The business was still not generating much cash, as few students were willing to pay to post their books to swap online. Charging at the big campuses caused many customers to stop using the website.

---

[8] "More on Mentoring," *Berry College, Campbell Columns*, (Fall 2003), 2.

Swarthout graduated in May 2004 during a poor job market. He moved into Pickering's home in the Atlanta suburbs. During this time, Pickering and Swarthout came up with the idea of PaperbackSwap.com. Pickering had amassed a large collection of books from extensive traveling and thought it would be nice if he could swap the books for different titles.

The Business Goes Live

PaperbackSwap.com was the first online book swap and used a systematic process, instead of forcing users to find someone offline with whom to trade. The site operated on a credit system and was free to use. The most difficult issue for PaperbackSwap.com to explain to members was paying for postage. The party sending the book was responsible for paying for postage. When members listed nine or more books, they received three free credits. After that, every time they mailed a book to a member, they received a credit to get a book. Also, members could receive additional credits by referring a person to the site.

Swarthout bought 600 books from Unclaimed Baggage Center (that is, unclaimed airline baggage) in Scottsboro, Alabama, and entered these into the system. (This was the initializing inventory to prime the book swap system.) The official launch date of PaperbackSwap.com was September 1, 2004. "In less than 9 months, the number of books had rocketed from 10,000 to more than 300,000 titles," according to co-founder Richard Pickering.[9] By October 2007, the site had more than 1,620,000 total books and 315,000 unique titles.[10] See Exhibit 12.1 for PaperbackSwap.com's book requests per day from the launch date until August 2007.

---

[9] "Paperbackswap.com Explodes," *The Glennville Sentinel*, June 15, 2006, PaperbackSwap.com, http://paperbackswap.com/press_media_detail.php?id=62.

[10] Robert Swarthout, PowerPoint presentation, October 19, 2007.

Competitors

Two weeks after PaperbackSwap.com opened, a copycat site (TitleTrader.com) launched. When PaperbackSwap.com added a feature, it would show up on this copycat site shortly afterward. When PaperbackSwap.com started, there were 10 competitors. However, if the sum of the second, third, fourth, and fifth competitors (in size) was doubled, PaperbackSwap.com was still larger.[11]

BookMooch

BookMooch.com was conceived, designed, constructed, and administered by John Buckman. As of February 1, 2008, BookMooch.com had 58,155 members.[12] As of February 1, 2010, the total number of books mooched was 523,427.[13] Unlike PaperbackSwap.com, BookMooch.com is a worldwide system. Sending to another country garners a shipper three points, and receiving books from another country costs two points. Each book entered into the BookMooch system gets the member one-tenth of a point. After members receive a book and leave feedback for the sender, they earn one-tenth of a point. The catch is that members have to send out at least one book for every two they receive. If a member does not maintain this ratio, he will no longer be able to mooch books, even if he has the points, until he improves his ratio. At BookMooch.com, members cannot purchase points, but they can contribute some of their points to "BookMooch-selected charities."[14] BoockMooch.com states that it pays its bills by using the Amazon affiliate program.[15]

---

[11] Robert Swarthout, personal interview, January 13, 2009.
[12] "Statistics About BookMooch," BookMooch website, http://bookmooch.com/about/stats.
[13] Ibid.
[14] Mary Pilon, "Online Swap Meets for Books," *Wall Street Journal*, August 7, 2008, http://online.wsj.com/article/SB121805999670918387.html.
[15] "Overview," BookMooch website, http://bookmooch.com/about/overview.

TitleTrader

Maggie Abercrombie and her brother Daniel Abercrombie created TitleTrader.com. Daniel handled the website construction, while Maggie provides customer service. TitleTrader.com of Millsboro, Delaware, allows users to receive and ship books domestically and internationally. TitleTrader.com gives members the option of using the basic free site or paying $19.95 for a premium membership. Premium features include wish list notification, auto-request discount, a newsletter, saved searches, as well as buddy lists. TitleTrader.com also offers games, DVDs, VHS tapes, CDs, and magazines. TitleTrader.com does not give members a free credit when listing their books. Members only receive credits when another member requests an item and that member receives it.

Business Model

PaperbackSwap.com's business model consists of a target market, revenue sources, and value configuration. PaperbackSwap.com has members who have become extremely loyal to this online enterprise.

Target Market

PaperbackSwap.com's target market is people with adequate disposable income and enough leisure time to read books. Its primary market remains women aged 40 and over who are typically not employed outside their residences.[16] The membership is comprised of 91% women.[17] These women have an abundant amount of time to read, and they generally read several books a week. They mainly swap erotic romance novels. When PaperbackSwap.com purchased books

---

[16] Robert Swarthout, personal interview, January 13, 2009.
[17] Ibid.

from the Unclaimed Baggage Center, Swarthout bought a broad assortment to appeal to a wide demographic. Shortly after the website went live, it became flooded with romance books.

By 2007, PaperbackSwap.com was covering all 50 states and 5 territories. As of February 2008, the company had approximately 100,000 subscribers, and approximately 66% would post the required 9 books to receive 3 free credits.[18] See Exhibit 12.2 for PaperbackSwap.com's map of unique member ZIP Codes at the time.

Revenue Sources

The original plan at PaperbackSwap.com was to start charging a membership fee at some point in the future, but this had yet to happen. PaperbackSwap.com used Amazon's affiliate program. When a book was not listed on PaperbackSwap.com, a member could click on the book to buy it from Amazon. Also, PaperbackSwap.com sold gift certificates and branded merchandise such as clothing, pens, mouse pads, and more. PaperbackSwap.com's revenue sources mainly came from selling credits, box-o-books, book journals, and postage.

Credits

The way PaperbackSwap.com initially generated cash was by selling credits. If members ran out of credits or had no books to list in the system, they could purchase a credit so they could order a book. This also allowed PaperbackSwap.com to reach people who did not want to list books or actively participate in the club, but just wanted a book for a low price. A credit initially cost $2.00 to purchase, and it cost $1.42 to mail a book.[19]

---

[18] Robert Swarthout, presentation, Campbell School of Business, Berry College, April 20, 2010.

[19] Robert Swarthout, personal interview, January 13, 2009.

Box of Books

In January 2006, PaperbackSwap.com introduced a feature called box-o-books for $8 a year. With box-o-books, a member could look at another member's list of available books (which was the only way to do this). The normal system was a queue that used a first-in, first-out system. With box-o-books, members did not have to use their credits. They just looked at another member's available books, and two members decided how many books they wanted to swap with each other. The two members had to agree on the number of books they wanted to swap because no credits were being used, and it was an even swap.

Book Journal

In November 2006, PaperbackSwap.com introduced the book journal feature. Like box-o-books, this feature also cost $8 a year. Book journal made it easy for members to organize their books by cataloguing them. It allowed members to assign books to spaces in their houses. For instance, if another member ordered a book, members would be able to see in what location of their house the book was located. This feature made it easier for members to keep their books organized and easier to respond to an order for one of their books.

Postage

PaperbackSwap.com calculated postage based on the weight of book. Book wrappers were printed by members on their computers. The books were wrapped in two sheets of paper because one sheet would usually tear. Ninety percent of the books could be wrapped in paper.[20] PaperbackSwap.com allowed members to print postage for a fee between $0.25 and $0.30, which allowed them to get delivery confirmation. With delivery conformation, PaperbackSwap.com showed the member where the book was traveling by placing a mail truck on

[20] Ibid.

a map. PaperbackSwap.com received data from delivery confirmation and could generally determine when a book would arrive. The company allowed the member to pay for just the delivery confirmation without paying for postage. Generally, a member received a credit when a book arrived at the shipping destination and the receiver marked that it had been received. Delivery confirmation allowed the member to get an instant credit instead of waiting for the book to arrive at its destination.

Value Proposition

PaperbackSwap.com offered a unique value proposition. The company's extreme dedication to customer service, mixed with the company's community atmosphere, created a website with loyal and dedicated members.

Customer Service

Almost all companies believe that customer service is part of their value proposition, but PaperbackSwap.com is extremely dedicated to customer service. For the initial two years of operation, Swarthout's own cell phone number was available on the website. He personally called members when they had a problem. He would also chat with members to see whether they liked the service and to identify potential improvements to be made. The goal was to answer communications within 24 hours.[21]

PaperbackSwap.com refused monetary donations from its members, even though it had many people willing to give money to help out the website. Instead, members donated their time to help operate the website. One member, Ruth H., began working 30 hours a week without compensation after being so impressed by Swarthout's personal call to her about a service problem with the website. Another member that Ruth H. had found, Linda S., was a medical doctor with

[21] Ibid.

sponge-like retention of information. She also began working without compensation. The company started using volunteer tour guides for the site, because it was important for people to learn how to use the system. Every member was assigned a tour guide. There were approximately 120 unpaid tour guides.[22] At times, it was easier to get in touch with a tour guide than paid PaperbackSwap.com personnel. A live chat was available if the member's tour guide was not online, although there was generally always a tour guide online in the chat area. Also, the company used volunteers to acquire book images and ISBN numbers.

To help get press coverage, PaperbackSwap.com mailed press releases with a book to local newspaper writers and received addresses for local newspapers from its members. Because PaperbackSwap.com had such a strong commitment to customer service, its customers became very loyal to the company.

Community

One of the main attractions for many of PaperbackSwap.com's customers is the community aspect. Members can create a profile and add members to their buddy list. A great deal of communication between members occurs in the discussion forums and the live chat. This has helped the company immensely, as social networking has increased tremendously in popularity.

The company also has fun competitions for members. Matt Stinnette won more than $300 worth of credits and shipping services from PaperbackSwap.com. He won by predicting the exact time the service would reach one million books traded. Based on figures determined by looking at daily trade averages, Matt predicted 10:38 a.m. on March 2, 2007.[23] His wife predicted the previous day.

[22] Ibid.

[23] Eric Feber, "Paperback Lover Hits Millionth Trade Jackpot," *The Virginian-Pilot*, March 16, 2007, PaperbackSwap.com., http://paperbackswap.com/press_media/press_media_detail.php?id=139.

Extendibility of Business Model

PaperbackSwap.com's model consists of a member ordering a book from the website, receiving the book through domestic mail, reading the book, and then possibly listing the book for another member to order. See Exhibit 12.3 for the PaperbackSwap.com model. Minimal modifications were needed to extend the PaperbackSwap.com model, mainly consisting of replacing books with another item. The model was capable of being extended into many areas, similar to what it had already accomplished with CDs and DVDs. The new sites were called SwapaCD.com and SwapaDVD.com, respectively. PaperbackSwap.com allowed members to transfer their credits between the three sites. In addition, the model could be used to swap video games, although a game cannot be sent via the media mail rate.

SwapaCD

On August 7, 2006, SwapaCD.com launched. SwapaCD.com followed the same basic concept as PaperbackSwap.com with some minor changes. With SwapaCD.com, members had to set up a SwapaCD money account in which they had to make deposits in $5.00 increments. Members used this money to pay a $0.49 transaction fee that was charged for every CD requested from another member.

SwapaDVD

On November 27, 2007, SwapaDVD.com launched. This site also followed the same basic model as PaperbackSwap.com. On the launch date at about 2:30 p.m., there were approximately 2,700 DVDs listed. By 9:30 p.m., about 6,700 DVDs were listed for swapping.[24] By February 2008, there were more than 36,000 DVDs available on the site. There were no transaction fees, unlike SwapaCD.com.

[24] Amy Brantley, "SwapaDVD Is Now Open," Associated Contest, November 27, 2007, Swapadvd.com, http://swapadvd.com/press_media/press_media_detail.php?id=25.

Financial Structure

PaperbackSwap.com is a privately held company, and therefore very little financial information is available. The company became cash flow-positive after nine months of operation. See Exhibit 12.4 for PaperbackSwap.com's approximate revenues for 2006. In the beginning, there were very few costs. The main expense was running the website utility, as no employee was receiving a salary. The company still has very limited costs, because of the small number of paid employees necessary to operate the website and perform other business functions. By 2007, the company only had seven full-time employees and five part-time employees.[25] PaperbackSwap.com owns no inventory, and the only costs are wages and salaries for website hosting, and so on. PaperbackSwap.com does not incur any expenses for shipping or mailing because that is the customers' responsibility. A prominent Silicon Valley venture capital firm offered to buy PaperbackSwap.com for $7.5 million, but the owners did not sell.[26]

Future Risks

Uncertainty with the future of media mail rates could destabilize PaperbackSwap.com's model. Club members might be unwilling to continue shipping books if the price of media mail continues to rise. Also, electronic book (e-book) readers are steadily gaining in popularity, which could be a serious threat to its business model because members cannot swap e-books. With the uncertainty of media mail and the gaining popularity of e-books, co-founder Robert Swarthout is faced with this difficult dilemma: The business enjoys continued growth, but in the future, such an online enterprise risks becoming

---

[25] Robert Swarthout, PowerPoint presentation, October 19, 2007.

[26] Robert Swarthout, presentation, Campbell School of Business, Berry College, April 20, 2010.

obsolete. With that future risk in mind, Swarthout ponders his professional fate with PaperbackSwap.com.

Robert is faced with the decision to remain at PaperbackSwap.com or leave. If he leaves, he must sell back his shares of the business at a relatively steep discount ("take a haircut"). This is required by a standard clause in the by-laws of startup companies such as this one. Robert would also have to agree to not work for a competitor or start another such venture; in other words, he would have to sign a noncompete clause in his buyout. Nevertheless, his settlement would be enough that he might not have to work for as many as 10 years with substantial annual income.

If Robert Swarthout stays at PaperbackSwap.com, several scenarios could develop. There was an unsolicited offer of $7.5 million to sell the business to a venture capital firm attempting to do an IT rollup. (Such a rollup occurs when a number of similar but independent technology firms are purchased, aggregated, and then sold as a package for significantly more than the combined purchase price of all the parts.) But the deal is contingent on Robert staying with PaperbackSwap.com for three years after the sale, as he is the IT expert at the firm. This type of contingency is not typical of acquisitions by venture capitalists, who usually dismiss the company founders in short order as they may not have the requisite skills for advancing the fortunes of the company that they have established.

A second possibility, assuming that Swarthout stays, is for PaperbackSwap.com to issue an IPO, but the offering is dependent on how well the business is doing against several competitors and whether the media mail rate, which their business model depends upon, is still intact. A third option is that PaperbackSwap.com could remain as a privately held company, with its continued success depending upon the strength of its competition, as well as the continued existence of the low media mail rate for books. But if Robert stays too long at what is becoming a boring situation, he might gain the reputation as a one-trick pony, capable of only "swapping" technologies. The relative probabilities of these three last scenarios, as well as the decision to go or stay, present a tough decision environment for Robert Swarthout.

Exhibits

Exhibit 12.1 Book Requests Per Day, September 2004–August 2007

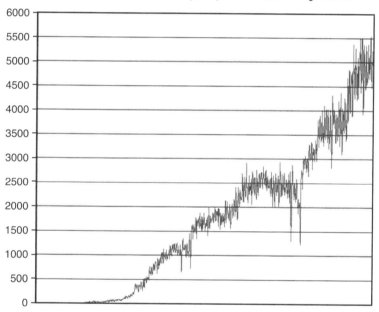

Source: Courtesy of Robert Swarthout

Exhibit 12.2 PaperbackSwap.com's Map of Unique Member ZIP Codes

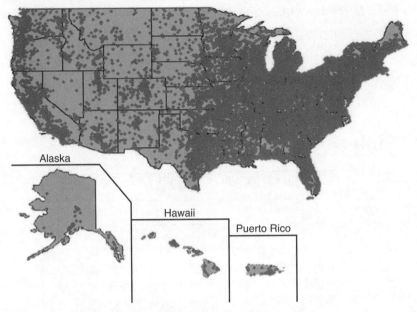

Source: Courtesy of Robert Swarthout

Exhibit 12.3 PaperbackSwap.com's Model

Source: Courtesy of Robert Swarthout

Exhibit 12.4 PaperbackSwap.com Approximate Revenues for 2006

Approximate Revenues	2006	Percent of Revenues
Delivery Confirmation	$40,000	12%
Box-O-Books	$10,000	3%
Book Journal	$24,000	7%
Branded Merchandise	$12,000	4%
Credits	$250,000	74%
Total	$336,000	100%

Source: Robert Swarthout, PowerPoint presentation, Berry College, November 2006

Case 13

Stranded in the Nyiri Desert: A Group Case Study

Aimée A. Kane and Mercy Shitemi,[1] Duquesne University

The Situation

You are part of a group of students participating in a study abroad program in Kenya. As a reward for exemplary performance, your group has been treated to a bus tour of the Nyiri Desert, which encompasses Amboseli National Park. It is the middle of August. You have been driving on a dirt road, far from the main road, to visit a local Maasai community. At about 11 a.m. the driver of your bus veers off the road to avoid hitting a wildebeest. The bus overturns, rolls into a ravine, and catches fire. The driver and the tour guide are both killed. Aside from some minor scratches and bruises, you and your fellow group members are not injured.

Several people in the city know your general itinerary, but due to the size and remoteness of Amboseli, they do not know your exact location, and you do not have any way of communicating with them. The closest settlement to where you have crashed is the Maasai community, which is approximately 40 miles to the south. When you do not arrive back at the hotel, people will realize something has happened, and the next day they are likely to look for you.

---

[1] Adapted from D. R. Johnson and F. P. Johnson, "Stranded in the Desert," *Joining Together: Group Theory and Group Skills*, 10th ed. (Upper Saddle River, NJ: Pearson, 2008). 332.

The Nyiri Desert is arid and, according to weather reports, temperatures could reach 92°F with a surface temperature of 100°F. You are all dressed in lightweight clothing and have hats and sunglasses.

As you escaped from the bus, each member of your group salvaged a few items, and there are 12 items in all. Your task is to rank these items according to their importance to your survival, from 1 (most important) to 12 (least important).

ITEMS	Individual Ranking	Group Ranking	Expert Ranking	Group Score[2]
Flashlight				
Pocket knife				
Amboseli Park map				
Large clear plastic ground cloth per person				
Compass				
Loaded gun				
Quart of water per person				
Book titled *Amboseli: Nothing Short of a Miracle*				
Heavy-duty canvas bus cover				
Matches				
Overcoat per person				
Side-view mirror				
			Total Group Score:	

Go to the next page to fill out and copy individual rankings to turn in.

---

[2] The group score is the absolute difference between the Group Ranking and the Expert Ranking. Lower scores are better as they indicate greater agreement with the expert.

Second Copy to Turn In

In the following, please copy your ranking from the previous page onto this page so that it can be turned in at the beginning of class session.

ITEMS	Individual Ranking
Flashlight	
Pocket knife	
Amboseli Park map	
Large clear plastic ground cloth per person	
Compass	
Loaded gun	
Quart of water per person	
Book titled *Amboseli: Nothing Short of a Miracle*	
Heavy duty canvas bus cover	
Matches	
Overcoat per person	
Side-view mirror	

4
Advanced Business Analytics

14. Joe's Coin Shop: Entry into Online Auctions   167
15. Vehicle Routing at Otto's Discount Brigade   181

Case 14

Joe's Coin Shop: Entry into Online Auctions

Charles A. Wood, Duquesne University

Introduction

Joe Murphy, founder of Joe's Coin Shop, sits back in his office chair to reflect on his 15-year-old business. It sure has been tough over the last decade, but the last couple of years have shown a profit. Joe thinks it is going to finally be okay, but he still wonders about the future direction of his company.

Company History

Joe remembered earning an MBA from a local college and working for a major corporation before deciding to go into business for himself. Joe thought that there could be nothing better than running a rare coin shop. Joe had been collecting coins since the fifth grade, and he knew all about how they are graded, what dealers typically offer for coins, and what they sell them for. It seemed like a natural fit. For the first five years, Joe had done really well for himself. He steadily improved sales and profitability up until his pinnacle in 2005, when Joe's Coin Shop had just broken $1 million in net income. Joe was making much more money than he made in the corporate world, and he was doing what he loved.

Joe took pride in providing for his customers—giving them the best advice, delivering a great selection, and allowing customers to purchase items on credit, which seemed to boost sales and encouraged customers to browse through his store. Unlike many stores that sold rare coins, Joe never took advantage of his customers' inexperience. He took great care to explain the difference between "blue book value," or the retail price, and "red book value," or the price a coin store should be willing to pay for the coin. If a customer did not know the value of a coin, Joe would examine it and be upfront about it, and truthfully describe the value rather than try to low-ball the estimate, as other dealers often do.

The customers appreciated Joe's honesty and friendliness. Many hung out at Joe's store on the weekends or every once in a while after work. Some brought him coffee or cookies, and they all sat around talking about coins, investing, and their individual collections.

But in 2006, sales just seemed to drop off. Customers who used to come to his store regularly simply stopped coming. In 2006, net income was reduced to one-eighth of the previous year, and during 2007–2009, the business lost more than $200,000 each year, despite cuts made to operating expenses and increases to liabilities. Joe just thought it was a momentary downturn, but he wondered how long he could hemorrhage money. Then, in 2010, at the height of a particularly bad recession, Joe's Coin Shop lost around $754,000. The record business profit from 2005 was wiped out during the 2007–2010 downturn. Business equity was down $330,000 from its high point, and times were bleak. Joe was worried that he would have to close the shop. Joe kept cash flow in the business by greatly increasing the debt load, but this could only go on for so long before Joe went bankrupt.

In late 2010, Joe decided to hire Anita Jones, a marketing consultant, to help him reach the customers who used to come to his store. Anita was certainly expensive, and Joe was worried about spending hundreds of thousands of dollars on a marketing campaign when times were so tough, but Anita was as good as she promised. Her marketing plan came through and helped bring the business back from the brink and into the black in 2012 and 2013. Joe is finally able to start paying down the liabilities that Joe's Coin Shop accrued during the 2006–2011 period. Business, however, is still not at the level that it once was, and Joe realizes that, while he now has some breathing room, something needs to be done if he is going to stay solvent over the long term. Exhibit 14.1 shows the sales revenue and net income from 2005–2013.

Market Survey

Anita surveyed past and present customers. In this market survey, Anita divided subjects into four groups:

- **Non-Customers**—Coin collectors (current or previous) who have never shopped at Joe's Coin Shop
- **Former Customers**—Coin collectors who have shopped at Joe's Coin Shop but have no plans to do so in the future
- **Infrequent Customers**—Coin collectors who shop at Joe's Coin Shop as well as other channels to buy rare coins
- **Loyal Customers**—Coin collectors who only shop at Joe's Coin Shop

The purpose of the survey was to determine why Joe's business was dropping off. Exhibit 14.2 shows the dominant answers to the questions from the market survey.

Anita and Joe discussed how Joe's Coin Shop is losing out to online coin stores and online auctions. Joe agrees that this might be the case. Joe has always contended that a traditional coin shop with great service and a personal touch would be more appealing to his clients, and indeed, some of them have expressed that exact thought.

Joe has so far resisted going online, and only has a simple website that talks about how he values the customer and gives the store's address and phone number, along with directions to the store. Nonetheless, Joe sees the writing on the wall. It is going to be more difficult to get customers in the future, and Anita has convinced him that he needs to start selling online to capture customers who are unwilling or unable to purchase inside his store.

But Joe really likes having a store, and he thinks that current customers enjoy it as well. He would hate to give that up. (Joe told Anita, "If I have to do all my work online, I would close up the business and go back to the corporate world!") Because Joe feels so strongly about this, Anita has suggested that Joe use the store to help. She suggested that he could advertise heavily within the Joe's Coin Shop store rather than covertly go online under another identity, and that a "bricks and mortar" store presence could add to the attraction of the website and to Joe's online auctions. Joe agrees. Joe also wants to make sure that he does not betray existing loyal customers by selling the same coins for less money in a different environment under an assumed name. Joe feels strongly that existing customers should be rewarded for staying with him, and not be denied opportunities that (hopefully) new customers will have, even if it costs him a little in the margin. Thus, existing customers should always be informed about new online coin collecting opportunities at Joe's Coin Shop.

Costs of an Online Market

Anita has suggested a two-pronged attack:

- First, an online storefront will sell items that are available at the store, but at the same price as the store. Depending on the level of interest, Anita encourages Joe to consider slightly lowering the prices on the site and at the store.

- Second, online auctions will allow Joe to reach a larger audience; they can be used as a vehicle to drive customers to Joe's site, where they can buy coins immediately rather than waiting for an auction to end. Some auction houses also allow customers to set up an online store presence within the auction house, so that if a customer wants to see auctions available from Joe's Coin Shop, she can simply go to the Joe's Coin Shop online store. Anita thinks that this is a good idea.

Joe's margins are at around 53% right now. If he were to sell in online auctions, he might increase the quantity of sales to previous levels, but at a reduced margin. Furthermore, both Joe and Anita feel that an online presence will eat into at least some of their sales as loyal customers purchase from them in online auctions. Anita has often said, "If we don't cannibalize ourselves, someone else will." Anita estimates that there will be a 25% migration from in-store to online, but also a 20% migration from the auction site to the Joe's Coin Shop's website, as some coin collectors do not want the uncertainty of the online auction.

Anita has explained that an online presence will require a website with online order capabilities, security, and transaction retrieval, which Anita suggests will cost around $200,000, plus an additional employee (estimated at $100,000 per year, including salary and benefits) to maintain the website, as well as enter information, keep the site active, and enter auction data while Joe deals with customers who enjoy the "bricks and mortar" store. In addition, auction participation will require additional listing fees for each item sold. Exhibit 14.3 shows a typical fee structure for an online auction.

Exhibit 14.3 shows a *listing fee* that depends on the starting bid. Then, if an item is successfully sold to a bidder, the seller owes a *sales fee* that consists of a fixed fee plus a percentage of the final bid amount. Finally, if a seller wants to place a secret reserve price on the coin for sale that the winning bid must surpass before the item is sold, a reserve fee is added. Listing fees and reserve fees are due immediately, whereas sales fees are due upon a successful sale.

Data Analysis

To help Joe with the planning, Anita has Joe purchase a data set of 2006 auction transactions. Analyzing this data will determine a proforma income statement (similar to a projected budget) to forecast the profits and expenses from the online information to discover how much income an online presence will deliver compared to the costs of implementing such a system.

The data looks daunting. There is so much data, separated into worksheets with rows in one worksheet pointing to several rows in the other worksheets. Anita and Joe identify five tasks for *each* question that need to be answered:

1. First, you must decide what questions you want answered. Remember, the questions should not be dependent upon the data set. Joe has several questions:
    a. What revenue can I expect to make compared to the book value of the coin? Do factors such as book value, number of bids, and so on, increase this revenue?
    b. Are there any factors that can help guide me to making more money per auction or to attracting more bidders to my auction?
    c. How many auction and non-auction coins do I need to sell online to break even with the cost of an online presence?
    d. Do bidders cluster into groups, and how can high-paying groups be targeted?
    e. How do losing bidders in one auction behave in the next auction when trying to buy the same item?

f. How does seller reputation affect sales? How does bidder reputation affect sales? Does the reputation score really measure the reputation?

2. Decide how you can answer the questions by using the data set.
3. Anita and Joe must take all this data and place it into a single data set for each question. Anita feels more comfortable staying with the spreadsheet, and has started coding VLookup commands and pivot tables to help with the analysis. Joe feels more comfortable with a database, and has migrated the spreadsheet data there and is using database views and queries, along with some SQL (the computer language of databases), to retrieve the data he needs.
4. Joe realizes that each coin is different from other coins, and there must be a way to standardize any data set. For instance, a 1797 penny in great condition worth $21,000 is very different from a 1909 penny in poor condition worth $15. If the 1909 penny sells for $18, that's much better than the 1797 penny selling for $20,098. There needs to be a way to compare expensive and cheap coins across the data set without simply saying that "this coin sold for more than that coin."
5. After a data set has been created for a specific question, that data set needs to be analyzed. Anita has informed Joe that simple averages are not enough, and will probably lead him down the wrong path, or leave some paths unexplored. Regression analysis, cluster analysis, and maybe even some time series analysis or panel studies must be in order. There are factors that can corrupt analysis, such as endogeneity or multicollinearity, that need to be considered, as well.

Joe is most concerned with measuring the critical success factors. He knows that he needs to measure and monitor online sales as they approach profitability, and he needs to know what that level will be and how far away from (or beyond) that level Joe's Coin Shop is right now.

Anita and Joe certainly have some work ahead of them.

Exhibits

Exhibit 14.1 Revenue and Income 2005–2013

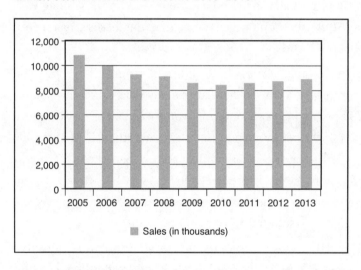

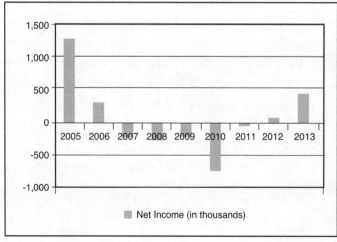

Exhibit 14.2 Main Responses to Multiple Categories of Questions

	Non-Customers	Former Customers	Infrequent Customers	Loyal Customers
Why do you/did you shop at Joe's Coin Shop?	N/A	It's just a place to get coins for my collection.	I like discussing coins. Nice service. It's just a place to get coins for my collection.	I like Joe. I like discussing coins. Nice service. Honest.
Why do you not or no longer shop at Joe's Coin Shop?	I only buy online. I can't afford to collect coins anymore.	Just too much hassle. It's friendly, but I like the in and out of online. I can't afford to collect coins anymore.	N/A	N/A
What do you look for in a coin shop?	Ease of access. Online availability.	Ease of access. Online availability.	Great service. Online availability. Knowledgeable staff. Safe environment/Joe doesn't try to rip me off.	Friendly service. Knowledgeable staff. Safe environment/Joe doesn't try to rip me off.
What would make you consider Joe's Coin Shop, continue shopping there, or increase the number of coins you purchase from Joe?	Online presence	Online presence	Online presence	Continued friendly service
When you purchase coins online, how much do you save?	45% (average)	28% (average)	25% (average)	N/A

Exhibit 14.3 Sample Seller Fees for an Auction Site

Starting Bid	Sales Fee	Listing Fee on Starting Bid	Reserve Fee
$0.01–$0.99	$0.00 + 5.25% of the winning bid	$0.25	$1.00
$1.00–$9.99		$0.35	
$10.00–$24.99		$0.60	
$25.00–$49.99	$0.625 + 2.75% of the winning bid	$1.20	
$50.00–$199.99		$2.40	$2.00
$200.00–$499.99		$3.60	1% of reserve price up to $100
$500.00–$1,000.00		$4.80	
>$1,000.00	$13.125 + 1.5% of the winning bid		

Exhibit 14.4 Joe's Coin Shop Balance Sheet

Balance Sheet (in thousands)			2013	2012	2011	2010	2009	2008	2007	2006	2005
Assets											
	Current Assets:										
		Cash and Cash Equivalents	4,551	4,511	4,514	4,477	4,670	4,522	4,382	4,274	4,331
		Accounts Receivable	1,587	1,587	1,587	1,587	1,587	1,587	1,660	1,872	2,052
		Inventories	461	461	461	461	461	461	464	437	432
		Total Current Assets	6,599	6,559	6,562	6,525	6,718	6,570	6,506	6,583	6,815
	Property and Equipment		3,774	3,377	3,310	3,392	3,879	4,208	4,494	4,610	4,252
	Other Assets		534	494	464	473	542	593	656	704	676
	Goodwill		5,530	5,350	5,550	5,140	5,670	5,620	5,760	6,470	7,120
	Total Assets		16,437	15,780	15,886	15,530	16,809	16,991	17,416	18,367	18,863

continued

Exhibit 14.4 continued

Balance Sheet (in thousands)			2013	2012	2011	2010	2009	2008	2007	2006	2005
Liabilities and Owner's Equity											
	Current Liabilities:											
		Accounts Payable	1,167	1,278	1,292	1,288	1,093	1,033	959	897	937
		Accrued Liabilities	2,836	3,105	3,139	3,130	2,655	2,509	2,330	2,178	2,275
		Accrued Compensation and Benefits	913	886	966	1,048	1,115	1,150	1,162	1,151	1,096
		Total Current Liabilities	4,916	5,269	5,397	5,466	4,863	4,692	4,451	4,226	4,308
	Deferred Rent		690	649	591	543	535	487	482	487	492
	Other Long-Term Liabilities		503	550	556	554	470	444	412	385	402
	Total Liabilities		6,109	6,468	6,544	6,563	5,868	5,623	5,345	5,098	5,202
	Owner's Equity		10,328	9,312	9,342	8,967	10,941	11,368	12,071	13,269	13,661
	Total Liabilities and Owner's Equity		16,437	15,780	15,886	15,530	16,809	16,991	17,416	18,367	18,863

Exhibit 14.5 Joe's Coin Shop Client Count

Client Count	2013	2012	2011	2010	2009	2008	2007	2006	2005
Returning Clients	482	503	464	512	510	522	561	645	691
New Clients	71	32	91	2	57	40	15	2	21
Total Clients	553	535	555	514	567	562	576	647	712

Exhibit 14.6 Joe's Coin Shop Statement of Operations

Statement of Operations (in thousands)	2013	2012	2011	2010	2009	2008	2007	2006	2005
Net Sales	$8,793	$8,694	$8,565	$8,430	$8,560	$9,096	$9,326	$9,986	$10,799
Cost of Goods Sold	4,431	4,510	4,203	4,709	4,529	4,711	4,407	4,766	5,097
Gross Profit	4,362	4,184	4,362	3,721	4,031	4,385	4,919	5,220	5,702
Sales and Marketing Expense	1,380	1,669	1,703	1,629	1,229	1,547	2,003	1,927	1,515
Accounts Payable Write-Off	102	88	110	82	99	97	100	112	149
Consulting Expense	50	40	20	5	-	-	-	-	-
Salary and Administration Expense	2,403	2,333	2,543	2,759	2,935	3,026	3,057	3,027	2,883
Total Operating Expenses	3,935	4,130	4,376	4,475	4,263	4,670	5,160	5,066	4,547
Net Operating Income (Loss)	$427	$54	($14)	($754)	($232)	($285)	($241)	$154	$1,155

Case 15

Vehicle Routing at Otto's Discount Brigade

Matthew J. Drake, Duquesne University

Introduction

Otto's Discount Brigade is a regional chain of discount stores mainly operating in southeast Ohio, with a few stores in West Virginia border towns. The firm operates a single distribution center in Cambridge, Ohio, located in close proximity to the intersection of Interstates 70 and 77. The distribution center has small refrigerated and frozen sections to accommodate the small number of stock-keeping units (SKUs) that require these special conditions.

Otto's Distribution Network

Every Otto's store is only slightly larger than an average, standalone drugstore; thus, it is not practical for the stores to receive shipments directly from suppliers because the shipments would require expensive less-than-truckload (LTL) transportation, as well as dedicated receiving personnel to handle the steady stream of shipments each day. Instead, the Otto's distribution center receives full truckload shipments from all suppliers and builds mixed pallets of products from many different suppliers to fulfill orders from the individual stores.

Stores submit their orders to the distribution center by 5 p.m. each day, and the distribution center picks, packs, and ships them by 11 p.m. Although the mix of products ordered by each store differs each day, the overall volume of each order (in number of mixed pallets) remains relatively constant. Exhibit 15.1 provides the typical daily demand requirements at each store.

Otto's operates a private fleet of three refrigerated trucks with drivers that can each drive for a maximum of 11 hours per night. Each truck can accommodate a maximum of 26 standard pallets, which means that delivering all the freight specified in Exhibit 15.1 requires at least one truck to travel on two delivery routes each evening. The locations of Otto's 15 stores, along with the distribution center, are depicted on the map in Exhibit 15.2. Exhibit 15.3 provides the typical travel times in minutes between all the locations in Otto's distribution network. It is safe to assume that the travel times between two locations are the same in either direction because the deliveries from the distribution centers to the stores occur in the late evening, when there is no traffic to speak of in rural southeastern Ohio. The lack of traffic also means that there is little variation in the travel times between locations from night to night; thus, these travel times can be assumed to be deterministic.

Otto's refrigerated trucks require approximately one gallon of fuel per hour to run the cooling units in addition to the fuel used to power the engine. With diesel prices near an all-time high and with no significant drop in fuel prices in sight, Otto's VP of Operations has challenged his logistics manager to consider re-establishing the delivery routes to reduce the total travel time. Of course, any travel time saved would also lower Otto's general fuel requirements (outside of the refrigeration units) and provide even more cost reduction to enhance the firm's profitability.

Currently, Otto's runs the following delivery routes each evening:

- **Route 1:** Cambridge–Zanesville–Pataskala–Mt. Gilead–Cambridge (25 pallets, 260 minutes)
- **Route 2:** Cambridge–Newark–Coshocton–Dover–Lisbon–Cambridge (24 pallets, 326 minutes)
- **Route 3:** Cambridge–Uhrichsville–East Palestine–Calcutta–Cambridge (24 pallets, 274 minutes)
- **Route 4:** Cambridge–Cadiz–Toronto–Weirton–Cambridge (21 pallets, 199 minutes)
- **Route 5:** Cambridge–Bellaire–New Martinsville–Cambridge (8 pallets, 195 minutes)

Route 2 is normally the only route that one truck will take in an evening. Another truck is assigned to Routes 1 and 4, and the third is assigned to Routes 3 and 5. The total travel time for all five routes is 1,254 minutes, or 20.9 hours per day.

Challenge for the Logistics Manager

In addition to reformulating the five delivery routes, the VP of Operations has also asked the logistics manager to see whether all of the daily deliveries can be covered by only four routes utilizing two of the three trucks (with a maximum travel time of 11 hours each). This additional consolidation would allow Otto's to redeploy one of the truck drivers as a warehouse worker (to replace a staff member who just moved out of the area), while keeping him available in reserve to make deliveries when store orders exceed their typical volumes.

After looking at the recent load reports and the vehicle efficiency statistics, the VP of Operations has the feeling that Otto's is overspending on its delivery operations. The problem is that he does not have the day-to-day insight into the operations to identify possible remedies. He has challenged the logistics manager to identify opportunities to improve the overall efficiency of the firm's logistics network. The stores and the distribution center are all part of the same company; thus, the stores would have to comply with any ordering directives put forward by the corporate logistics department.

Exhibits

Exhibit 15.1 Daily Demand Requirements (in Pallets) for Each Otto's Store Location

Location	Pallets per Day
Bellaire, OH	5
Cadiz, OH	5
Calcutta, OH	9
Coshocton, OH	4
Dover, OH	6
East Palestine, OH	10
Lisbon, OH	8
Mt. Gilead, OH	9
New Martinsville, WV	3
Newark, OH	6
Pataskala, OH	10
Toronto, OH	9
Uhrichsville, OH	5
Weirton, WV	7
Zanesville, OH	6

Exhibit 15.2 Map of Otto's Store Locations and the Cambridge, OH Distribution Center

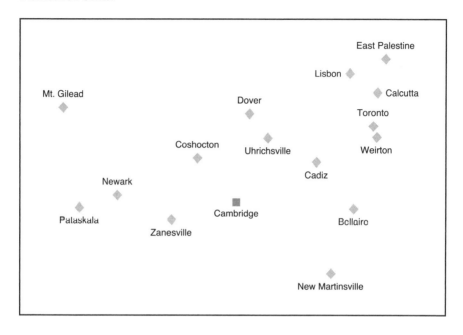

Exhibit 15.3 Travel Times (in Minutes) Between Each Otto's Store Location and the Distribution Center

Travel Times (in Min)	Cambridge, OH	Bellaire, OH	Cadiz, OH	Calcutta, OH	Coshocton, OH	Dover, OH	East Palestine, OH	Lisbon, OH
Cambridge, OH	0							
Bellaire, OH	56	0						
Cadiz, OH	56	46	0					
Calcutta, OH	108	70	56	0				
Coshocton, OH	47	94	70	119	0			
Dover, OH	45	88	47	86	46	0		
East Palestine, OH	129	93	79	23	130	90	0	
Lisbon, OH	106	85	71	18	107	68	24	0
Mt. Gilead, OH	119	161	163	153	83	109	152	137
New Martinsville, WV	94	45	82	106	128	125	127	117
Newark, OH	62	105	105	154	44	81	165	142
Pataskala, OH	67	110	115	164	65	101	185	163
Toronto, OH	90	52	38	27	104	81	48	38
Uhrichsville, OH	43	77	33	82	43	20	100	78
Weirton, WV	84	46	32	33	98	75	55	44
Zanesville, OH	33	75	80	130	48	67	151	129

Case 15 • Vehicle Routing at Otto's Discount Brigade

Travel Times (in Min)	Mt. Gilead, OH	New Martinsville, WV	Newark, OH	Pataskala, OH	Toronto, OH	Uhrichsville, OH	Weirton, WV	Zanesville, OH
Cambridge, OH								
Bellaire, OH								
Cadiz, OH								
Calcutta, OH								
Coshocton, OH								
Dover, OH								
East Palestine, OH								
Lisbon, OH								
Mt. Gilead, OH	0							
New Martinsville, WV	198	0						
Newark, OH	72	142	0					
Pataskala, OH	63	147	27	0				
Toronto, OH	173	90	140	150	0			
Uhrichsville, OH	119	114	80	99	65	0		
Weirton, WV	174	83	134	144	21	60	0	
Zanesville, OH	93	112	37	45	112	65	106	0

Index

A

AAD (Assistant Art Director), AGP Publishing Company, 86
ABS (acrylonitrile butadiene styrene) plastic laptop cases, 113, 115
   life-cycle analysis, 116
AD (Art Director), AGP Publishing Company, 86
advertising
   direct mail, 19
   email marketing, 20
   FeedMyPet.com, 17, 19-20
   magazines/newspapers, 19
   PPC (pay per click), 20
   SEO (search engine optimization), 20
   telemarketing, 20
   television, 19
AE (Assistant Editor), AGP Publishing Company, 85
aluminum
   availability, 113
   greenhouse data, 129
   industry information, 108-109
   laptop cases, compared to plastic, 113-116
   laptops, 107-108
   LCA data, 116
   production, 114
ANOVA (analysis of variance), Pizza Station, 39

APG Publishing Company
    background, 84-85
    history, 84
    income statements,
        samples, 97
    independent contractors, 90
        hiring decisions, 93
    outside resources, 90
    personnel availability, 96
    printers, 88
    production department,
        85-86
    production process, 86-89
    production schedule, 89-92
        budget buffer, 92
        *general processing
            times, 95*
        problems in, 91
        sales potential and, 91
    terminology, 94
assets, value increase, 18
average growth rate, 8-9

B

backordering
    costs, 52
    measuring, 50
BBLS (barrels sold), 5
BookMooch.com, 148
breweries
    NBC (Narragansett Brewing
        Company), history, 101-103
    questions, 6-10

budget
    buffer, publishing
        company, 92
    inventory management
        and, 51
    overspending, 51
    sales forecasting and, 11
    St. Elizabeth Seton Catholic
        Church, 27-30
**business opportunities,
    starting, 146**

C

CAGR (compound annual
    growth rate), 109
cash flow, St. Elizabeth Seton
    Catholic Church, 27-29
CE (consumer electronics)
    Durable Aluminum Inc., 118
    recycling, 112-113
    take-back programs, 128
change, resistance to, 51
collegiate book swap. *See*
    PaperbackSwap.com
Commonwealth Pipeline
    Company
    background, 77-78
    connections and distance
        between terminals, 77
    facilities, 79
**communication, Durable
    Aluminum Inc., 123**

computer hardware
  Durable Aluminum Inc. and, 111
    market entry, 111-112
  industry sales, 111
consumers, Durable Aluminum Inc., 117-119, 123
continuous order systems, 52
control chart, order errors (Pizza Station), 39
COO (Chief Operations Officer), AGP Publishing Company, 85
cost analysis, inventory, 68-69
craft brewing, 4, 6
CTQ (critical to quality), 35
cumulative distribution function, 67
Current State Map, Pizza Station, 39
curves, 7
customer feedback, 34
customer satisfaction study, 35
customers, VOC (voice of the customer), 35

D

data analysis, methods, 173
databases, data analysis and, 173
Decision Science, 93

design, Durable Aluminum Inc., integration of aluminum, 112
direct mail advertising, 19
distribution function, cumulative, 67
Durable Aluminum Inc.
  aluminum *versus* plastic laptop cases, 113-116
  background, 109-110
  case challenges, 125
  computer hardware industry and, 111
    market entry, 111-112
  income statement, 110
  Laptop Strategic Planning Team, 116
    communication strategy, 123
    consumers, 117-119
    environmental impact, 120-121
    pricing, 121-122
    product lines, 119-120
    profitability, 121-122
  laptops, 107-108
  New Business Development Initiative, 116
  overview, 107-108
  product design, integration of aluminum, 112
  recycling program, 123-124
  sustainability, 110
  waste minimization, 110

E

email marketing, 20
environment
   aluminum greenhouse data, 129
   Durable Aluminum Inc., 120-121
      enhanced recycling program, 123-124
      recycling and, 122
   plastic greenhouse data, 130
e-waste, recycling and, 112-113

F

fashion retail, inventory manage-ment, 65
feedback, 34
FeedMyPet.com
   advertising campaign, 17
   assets increase, 18
   financial success, maintenance, 18-19
   IPO, 15-16
   marketing, 17, 19-20
   operating margins, 19
   overview, 15-17
financial performance measures, 50
forecasting
   benefits, 11
   offertory revenue, 25
   Ska Brewing Company, 3, 6-9
fractional charge per short unit, 52
fundraising, St. Elizabeth Seton Catholic Church, 29
Future State Map, Pizza Station, 40

G

Global Supply Chain Operations, 48
   financial performance measures, 50
   inventory management, 49, 52
      financials and, 50
      order systems, 52
      SKUs, 50
   SKU, 50
greenhouse data
   aluminum, 129
   plastic, 130
growth
   average growth rate, 8-9
   median growth rate, 8-9
   percentage growth, 8

H

HealthCare
   balance sheet, 134-141
   Corporate Social Responsibility program, 135

energy saving project
  assumptions, 140-137
history, 133-134
HVAC systems, 136
  expenses, 137
income statement, 139-134
interior tile replacement,
  137-138
  expenses, 142-138
mission statement, 139
sustainability, coordinator,
  131-133
tree replacement, 136-137
  expenses, 137
vision, 134
website information, 139
HVAC (heating, ventilation, and air conditioning)
system, 136

I

implementation plans, Pizza
  Station, 40
income statements, Durable
  Aluminum Inc., 110
independent contractors,
  publishing company, 90
  hiring decisions, 93
inventory cost analysis, 68-69
inventory management
  backorder costs, 52
  backordering, 50

budget and, 51
cumulative distribution
  function,
daily demand, standard
  deviation, 66, 73
fashion retail, 65
financials and, 50
fractional charge per short
  unit, 52
Global Supply Chain
  Operations, 49
holding costs, 52
order systems, 52
overstock, safety stock levels
  and, 68
replenishment lead time, 67
safety stock formula, 69
safety stock level, 66, 67
  overstock and, 68
saving, quantity, 68
savings, 68
service targets, 67
shortage cost, 52
SKUs, 50
stockout, 52
target inventory service level,
  66-67, 75
transaction costs, 52
IPO (initial public offering)
  FeedMyPet.com, 15-16
  PaperbackSwap.com, 157
IT rollup, 156

J-K

Joe's Coin Shop, 167
　balance sheet, 178
　data analysis, 172-174
　　client count, 174
　　main question responses, 175
　history, 167-169
　market survey, 169-170
　online auctions
　　market costs, 170-172
　　seller fees samples, 171
　online auctions and, 170
　statement of operations, 174
JTMC
　inventory levels, 66
　　daily demand, 66
　　target inventory service level, 75
　inventory management, 66-69
　overview, 65-66
　safety stock, 67

L

laptops, aluminum cases *versus* plastic, 113-116
LCA (life-cycle analysis), ABS (acrylonitrile butadiene styrene), 116
Lean Principles, 35

M

MAD (mean absolute deviation), 8
magazine/newspaper advertising, 19
MAPE (mean absolute percentage error), 8
marketing
　direct mail, 19
　email marketing, 20
　FeedMyPet.com, 17, 19-20
　magazines/newspapers, 19
　PPC (pay per click), 20
　SEO (search engine optimization), 20
　telemarketing, 20
　television, 19
ME (Managing Editor), AGP Publishing Company, 85
median growth rate, 8-9
monthly data, Ska Brewing Company, 9-11

N

NBC (Narragansett Brewing Company)
　community fit, 104
　distribution center, 103-106
　history, 101-103
　keg facility, location decision, 103-106

operating costs, location and, 104
space needs, 103-106
Nyiri Desert trip, 161-163

O

offertory revenue forecasting, 25-27
  past data, 30
online market costs, 170-172
operating costs
  FeedMyPet.com, 19
  NBC (Narragansett Brewing Company), location and, 104
order systems
  continuous, 52
  periodic review, 52
Otto's Discount Brigade, 181
  daily demand requirements, 184-182
  delivery, 182-183
    logistics, 183
    spending, 184
    travel times, 182-187
  distribution, 181-183
  locations map, 182
  overview, 181
  refrigerated trucks, 182
outliers, growth rates, 8-9
overspending, 51
overstock, safety stock levels and, 68

P

PaperbackSwap.com, 143, 146
  book requests per day, 147
  BookMooch.com, 148
  business model, 149-153
    extendibility, 154
  credits, 147
  financial structure, 155
  future risks, 155-157
  history, 143
  initial operations, 146-147
  launch date, 147
  map of unique zip codes, 150
  model, 154
  origins, 144-145
  postage payment, 147
  revenue sources, 150
    book journal, 151
    box-o-books, 151
    credits, 150
    postage, 151-152
  revenues for 2006, 155
  SwapaCD.com, 154
  SwapaDVD.com, 154
  target market, 149-150
  TitleTrader.com, 149
  Unclaimed Baggage Center, 147
  value proposition, 152
    community, 153
    customer service, 152-153
percentage growth, 8

performance, financial performance measures, 50
periodic review order systems, 52
Pizza Station
　analysis, 38-40
　assembly system, 36-37
　background, 33-36
　baking system, 37
　boxing system, 37-38
　cutting system, 37-38
　delivery system, 38
　labeling system, 37-38
　ordering system, 36
　storage, 38
　supplies, 38
plastic
　ABS (acrylonitrile butadiene styrene), 115
　　life-cycle analysis, 116
　greenhouse data, 130
　production, 114-115
PM (Production Manager), AGP Publishing Company, 86
PPC (pay per click), marketing and, 20
printers, APG Publishing Company, 88
process, VOP (voice of the process), 35

product design, Durable Aluminum Inc., integration of aluminum, 112
production schedule, publishing company, 89-92
publishing firm
　See also APG Publishing Company
　industry overview, 82-83
　introduction, 81-82
　terminology, 94

Q

quality, CTQ (critical to quality), 35

R

recycling
　CE (consumer electronics), 112-113
　Durable Aluminum Inc., 122
　　enhanced recycling program, 123-124
replenishment lead time, 67
resistance to change, 51

S

safety stock level, 69, 67
　target inventory service level and, 66

sales, production schedule and, publishing company, 91
sales forecasting
  benefits, 11
  Ska Brewing Company, 3
    annual data, 6-9
    monthly data, 9-11
scatter plots, 6-7
  sales forecasting, 6-7
schedules, production schedule, publishing, 89-92
SEO (search engine optimization), marketing and, 20
shortage cost, 52
Ska Brewing Company, 3
  annual data, 6-9
  background, 3-4
  BBLS (barrels sold), 5
  mission, 5
  monthly data, 9-11
SKU, 50
spreadsheets, data analysis and, 173
St. Elizabeth Seton Catholic Church, 25
  cash flow analysis, 27-29
  fundraising, 29
  history, 26-27
  overview, 25-26
  past data, 30

Steamworks Brewing Company, 4
stockout, 52
sustainability, 118
  aluminum greenhouse data, 129
  Durable Aluminum Inc., 110, 118
  HealthCare, 131
  plastic greenhouse data, 130
SwapaCD.com, 154
SwapaDVD.com, 154

T

Takt time, Pizza Station ordering, 39
target inventory service level, 66-67, 75
  overstock and, 68
telemarketing, 20
television advertising, 19
tile replacement, 137-138
TitleTrader.com, 149
tree replacement, 136-137
trendlines, 7

U

Unclaimed Baggage Center, PaperbackSwap.com and, 147
used books, 143. *See also* PaperbackSwap.com

V

venture capitalists, IT
   rollup, 156
VOC (voice of the
   customer), 35
VOP (voice of the process), 35

W

WM (Warehouse Manager),
   AGP Publishing Company, 86

X-Y-Z

X-bar-R chart, Pizza Station
   pizza thickness, 39